The Joy of Science

We live in an age where working in science or engineering offers tremendous professional opportunities – the pace of scientific development is truly breathtaking. Yet many researchers struggle with the pressures of the fast-paced academic workplace, and struggle to harmonize their work and personal lives. The result can be burnout, exhaustion, and stress on a personal level, and difficulty in recruiting and retaining talented, diverse people in science and engineering. This book, written for graduate students and researchers at all stages of their careers, aims to help scientists by identifying and questioning the core beliefs that drive a culture of overwork, and provides real-world examples and exercises for those wishing to do things differently. Written in a lively, narrative style, and including interview excerpts from practicing scientists, social scientists, and engineers, this book serves as a guide for those seeking to practice the seven traits of the joyful scientist.

ROEL SNIEDER holds the Keck Foundation Endowed Chair of Basic Exploration Science at the Colorado School of Mines. He received a master's degree in geophysical fluid dynamics from Princeton University in 1984, and in 1987 a PhD in seismology from Utrecht University. He has a strong interest in teaching professional development, and has developed classes on the Art of Science, Research Ethics, and Teamwork and Leadership. He is coauthor of the textbooks *A Guided Tour of Mathematical Methods for the Physical Sciences* (Third Edition, 2015; Cambridge University Press) and *The Art of Being a Scientist: A Guide for Graduate Students and their Mentors* (2009; Cambridge University Press). Roel is a Fellow of the American Geophysical Union and Honorary Member of the Society of Exploration Geophysicists.

JEN SCHNEIDER is an associate professor in the Department of Public Policy and Administration at Boise State University, where she teaches and advises in the graduate program, and holds affiliate appointments with the College of Innovation and Design and the Public Policy Research Center. She serves as the associate director of the Boise State University Experimental Program to Stimulate Competitive Research (EPSCoR) program. She has published widely in the fields of environmental and science communication, energy studies, and engineering studies, including coauthoring the books *Engineering and Sustainable Community Development* (2010) and *Under Pressure: Coal Industry Rhetoric and Neoliberalism* (2016).

"In these times where scientists are under increasing pressure to prove their worth through metrics, this welcome book provides a refreshing perspective into how we might all find personal satisfaction and joy in playing the game of science."

– Julie Smith, Psychologist, Radford College, and Malcolm Sambridge,
Australian National University

"This book stands out from the sea of recipes for work-life balance and time management, in that it distills joy as a core trait for "success". Although it may seem at odds with the objective scientific method and the image of detached, humorless scientists, the authors make a fascinating case for personal touch, passion, and subjective imprint as core assets for creative, ground-breaking progress and improved personal lives. The book is informative and captivating, but most importantly, it is timely in an age where performance pressures surge to the detriment of progress, and academics find themselves with little time to do research."

– Tarje Nissen-Meyer, University of Oxford

"Too many scientists fall into the trap of overworking and burning out at great cost to their careers and personal lives. In *The Joy of Science*, Roel Snieder and Jen Schneider present a highly readable discussion of the challenges scientists face, and provide personal stories, thought-provoking questions, and practical recommendations relevant to both young and more senior scientists. They remind us that we are human and help us keep our eye on the ball, of a joyful and fulfilling life."

– Richard Primack, Boston University and Editor-in-Chief,
Biological Conservation

"In *The Joy of Science*, emotional intelligence takes its rightful place alongside intellectual intelligence. A perfect read for the scientist, academic or engineer seeking a more fulfilling and successful life. [It] will be at the top of my client reading list."

– Eileen Flanigan, MBA Process Engineer, Author and Life Coach

The Joy of Science

Seven Principles for Scientists Seeking Happiness, Harmony, and Success

ROEL SNIEDER
Center for Wave Phenomena
Colorado School of Mines

JEN SCHNEIDER
Department of Public Policy and Administration
Boise State University

CAMBRIDGE
UNIVERSITY PRESS

CAMBRIDGE
UNIVERSITY PRESS

University Printing House, Cambridge CB2 8BS, United Kingdom

One Liberty Plaza, 20th Floor, New York, NY 10006, USA

477 Williamstown Road, Port Melbourne, VIC 3207, Australia

314-321, 3rd Floor, Plot 3, Splendor Forum, Jasola District Centre, New Delhi - 110025, India

79 Anson Road, #06-04/06, Singapore 079906

Cambridge University Press is part of the University of Cambridge.

It furthers the University's mission by disseminating knowledge in the pursuit of education, learning and research at the highest international levels of excellence.

www.cambridge.org
Information on this title: www.cambridge.org/9781316509005

© Roel Snieder and Jen Schneider 2016

First published 2016

A catalogue record for this publication is available from the British Library

Library of Congress Cataloging in Publication data
Snieder, Roel, 1958– | Schneider, Jennifer J.
The joy of science : seven principles for scientists seeking happiness, harmony, and success / Roel Snieder, Center for Wave Phenomena, Colorado School of Mines, Jen Schneider, Department of Public Policy and Administration, Boise State University.
Cambridge : Cambridge University Press, 2016. | Includes bibliographical references.
LCCN 2016006104 | ISBN 9781107145559
LCSH: Creative ability in science. | Science – Vocational guidance. | Science – Psychological aspects. | Engineering – Vocational guidance. | Engineering – Psychological aspects. | Job satisfaction.
LCC Q172.5.C74 S595 2016 | DDC 502.3–dc23
LC record available at http://lccn.loc.gov/2016006104

ISBN 978-1-107-14555-9 Hardback
ISBN 978-1-316-50900-5 Paperback

For our families
Eric, Addie, and Nolie,
Idske, Hylke, Hidde, and Julia,
who make seeking harmony worthwhile.

And for Terry Young , a great friend and colleague,
for setting an example by embodying
wisdom and compassion.

Contents

Preface

The idea for this book came out of the experience of working with the faculty, staff, and graduate student reading groups at the Colorado School of Mines (CSM), in Golden, Colorado. Organized by Roel, these groups began meeting in 2009, and functioned as a sort of academic "book club," the aim of which was to provide our CSM colleagues with a venue for discussing interesting ideas and concepts related to working at the university. For example, one early text the group discussed was John Medina's *Brain Rules*, which discussed novel and innovative ways to promote learning and productivity.

But a secondary outcome soon emerged – as the two of us participated in the groups, we noticed that they were meeting a need among faculty, staff, and students to build a community, a place where we could discuss ideas that were meaningful to us not just as employees but as whole people, humans who lived their lives through and beyond the walls of the institution. We read books and readings on topics ranging from spirituality to social justice to project management, and were edified by our colleagues' commitment to showing up for sometimes difficult but often rewarding conversations about how to bring our whole selves into the workplace.

We also realized, however, that there were few books that addressed the specific concerns of scientists, social scientists, and engineers who wanted to better integrate their home and work lives, or who were struggling to feel both successful and joyful at work. In off-handed terms over lunchtime conversations, the two of us began discussing the possibility of creating a workbook for our reading groups that might offer helpful exercises and short readings for addressing the concerns of those trying to navigate their way through scientific or technical graduate programs, the tenure-track,

Janwillem Snieder (www.jwsnieder.nl)

or demanding academic and corporate careers in the sciences
that seem to require ever more commitment and longer working
hours.

As our discussions continued, we soon realized that a brief
workbook wouldn't allow us to do what we really wanted to do, which
was to describe the traits of those whom we believed were most
successfully and joyfully integrating their work and personal lives.
We noticed that the colleagues, mentors, students, and friends who
most inspired us shared seven characteristics, traits, or practices that
made them both effective and joyful in the workplace and beyond.
Exploring these seven traits required a more sustained exploration,
and the result of that exploration is this book. While working on this

project, we also had the opportunity to reflect on our own experiences, on reading we have done as "seekers" interested in untangling the puzzle of "work-life balance," and on some of the academic research that has been published on this topic. We have tried to integrate these many pieces into an easily accessible format, one that will primarily appeal to academic scientists, engineers, and social scientists, but which may have broad appeal for any professionals seeking more joy in their lives.

As with any book-length project such as this one, many hands made the work possible. We are grateful to Matt Lloyd and Zoe Pruce at Cambridge University Press, and Ramya Ranganathan at Integra-PDY, for their support, enthusiasm, and guidance in shepherding this book from draft to physical artifact. Sara Kate Heukerott, Christoph Sens-Schönfelder, and Ken Larner read early drafts of the manuscript and provided valuable feedback and encouragement. Rembrandt Zuijderhoudt was both an inspiration and a sounding board when this project was no more than an idea. Janwillem Snieder, Roel's brother, graciously provided the delightful illustrations for the book. We have also both had wonderful mentors, teachers, and role models who have helped us personally to live more joyful professional lives, and who were our exemplars as we brainstormed the seven traits of the joyful, successful academic: Ken Larner, Toni Lefton, Carl Mitcham, and Terry Young.

Several colleagues took valuable time at the end of a very busy semester to provide us with narratives of their own experiences – both good and bad – with integrating their work and home lives. We are incredibly grateful to them for their bravery and honesty, and their stories have made this book much richer. We are also lucky to have worked with many of them: Deserai Anderson Crow, Jason Delborne, Lejo Flores, Monica Hubbard, Ken Larner, Jerry Schuster, Jon Sheiman, Evert Slob, Sven Treitel, and Terry Young.

Above all, we are grateful for our families. Roel is grateful for Idske for sharing her love and for living with a restless husband, and for Hylke, Hidde, and Julia for being wonderful adult friends. Jen wants to thank Eric, who shows his support and love through countless gestures large and small, every day, and Addie and Nolie, who have made life better, richer, and more beautiful in all ways.

WE WOULD LIKE TO HEAR FROM YOU

Writing this book, and using it in our teaching, is an evolving project. We see this book as an exciting step in our own growth, and hope that this project leads to new initiatives such as lectures, workshops, or even another book. We would love to hear your reactions and suggestions, and hope you feel free to contact Roel (rsnieder@mines.edu) or Jen (jenschneider@boisestate.edu). We will post errata and other information related to the book, as well as the next steps that we will take, on the following website: www.mines.edu/~rsnieder/Joy_of _Science.html. Follow us on Facebook at www.facebook.com/Science HarmonySuccess/.

Introduction

Success is where you have found your joy.

–From the movie *Papadopoulos and Sons*

Sometimes, as young academics, we get advice about how to play the "game" of science, meaning there are rules of the game, both explicit and implicit, that we must master to succeed. Used in a cynical way, the metaphor of the "game" implies that things might be rigged against us, and that there are rules we have to learn or manipulate to succeed. Indeed, there are a number of tricks of the trade – systematic ways to respond to a journal editor's "revise and resubmit" decision, or how to compile a tenure dossier – that can make your success as an academic or scientist more likely.

But this book isn't about such tricks.

In fact, we – Roel and Jen – think the "science as game" metaphor can be useful. But thinking about science as a rigged game in which you must always be on your guard, where you are going to be chewed up and spit out by savvier "players," or where you are stuck playing one stultifying role is of limited usefulness. Instead, we like to think of science as a "game" in its most playful sense, one that invites exploration and venturing into the unknown as "moves," and discovery as the "prize." We believe there is an inherent playfulness in the practice of science, and it is probably this playfulness that drew many of us to scientific inquiry in the first place. At its very best, "doing" science is an activity that matches our innate drive to learn and explore new territory. We believe the most successful, joyful scientists are those who are able to keep this spirit of play, even as they also work hard and maintain vibrant personal lives.

There is an inherent playfulness in the practice of science.

This book is about figuring out how to maintain this sense of playfulness and joy as a scientist in the face of pressures to "play the game" in less meaningful ways. Messages about how to succeed as academics, or about the sorry state of the university system today, are often negative or demoralizing, and can leave new faculty members bewildered about what to expect or how to feel and act. We do offer practical advice for success here, but this advice is not presented in a bulleted list of how-tos. Instead, our hope is that readers will use this book to identify and establish core values that will lead them to success from a deeper place of significance, satisfaction, and meaning-making.

The thrill of discovering something new is enormous, and it can lead to recognition and advancement. We often push our graduate students and junior colleagues to focus on how they can make a "contribution," and the mass media and our university public relations offices are certainly very keen on scientific "breakthroughs." Making a new scientific discovery not only provides intellectual satisfaction, but also provides recognition, career opportunities, and possibly economic gain. In addition, scientific discoveries can be useful, which, in itself, is a satisfying reason to be a scientist. Scientific inquiry can help us to get a better grip on the world around us, and the discoveries of science have without a doubt shaped the world in myriad ways, from the introduction of drugs and treatments to cure diseases, to space exploration, to natural resource development, to the information revolution. One might view particular scientific endeavors with caution or delight, but it is hard to deny that science has shaped our material world. With modern instrumentation and dissemination of information, science has influenced the way in which we observe, understand, and experience the world, largely emphasizing the importance of data or objectivity in decision-making and policy.

A corollary to this objectivity, however, is that it may lead scientists to promote a purely mechanistic view of the world where only things that can be measured are considered to be real. This mechanistic view limits our abilities to interact with one another in satisfying ways. Addressing this limitation is one aim of our book.

We also know that scientists are a trusted source of information. Despite the evidence that there is a "war on science" in the United States (Mooney, 2005; Oreskes & Conway, 2011) or that Americans seriously lack scientific literacy (Mooney & Kirshenbaum, 2010) many Americans still rank scientists very high on their list of trusted sources of information (Gauchat, 2012). Even given the prevailing stereotypes of the "mad scientist" (Frayling, 2005) going into scientific fields remains a respected career path, and one that is potentially financially rewarding as well. For many, therefore, becoming a scientist is something to aspire to. Books celebrating the activity of scientific research – such as an earlier book also titled *The Joy of Science* (Sindermann, 1985) and Barbara Minsker's *The Joyful Professor* (2010) – are useful resources. Our work hopes to build on this earlier work by emphasizing not just the practice of science, nor the science of time management, but by helping us to articulate and then construct professional and personal lives that are in harmony and bring us joy.

Some scientists are already living harmonious, joyful, and successful lives. If this describes you, bravo! However, we believe that for many young scientists – and even for some of us who are more experienced – this somewhat romantic image belies the reality. In fact, many scientists live under enormous pressure. There is the pressure to produce scientific papers, encapsulated by the common wisdom to "publish or perish." There is the pressure to have a vocal presence at scientific meetings and to participate in committees and editorial boards. And for those in the academic community, there is the pressure to teach well, in addition to being innovative and productive in research.

Furthermore, many scientists view science as an activity that is inherently competitive. And there is indeed a competition to be the first to make a discovery, as there is pressure in acquiring research funding and job opportunities. These pressures, whether real or perceived, can be so large that the "joy of science" seemingly degenerates into the "survival of the fittest." These pressures can be aggravated by the expectation that in addition to having a successful career, we should also have a healthy and rich personal life. That personal life may involve raising a family with two working parents, children, and/or aging parents or loved ones who depend on us. And this says nothing of time needed to maintain or improve one's physical, mental, and even spiritual health. Trying to figure out how to make both professional and personal lives "work" puts an additional pressure on scientists, especially in the early stages of their careers.

But there is an even more insidious aspect to the pressure that many scientists feel, which is the commonly held belief that no matter how hard we work, *it is never enough*. Or perhaps we feel that *we* are never enough. No matter how many papers one

Working under the commonly held belief that no matter how hard we work, it is
never enough.

might have written, one can always write more. Even though one may have attended many scientific conferences, there are always more meetings to attend, and there are always more committees and editorial boards one can serve on. There are more grants to secure and more students to graduate. To make matters worse, the metrics seem to be changing, workloads are increasing, and sometimes the resources we need to do our work diminish. What further drives this feeling of pressure is the common belief among scientists that to be useful it is necessary to be "the best." This belief is often fueled in the formative years of graduate school by advisors who, often with the best intentions, want their advisees to be productive and shine in the scientific community. To achieve this, advisors sometimes push their students ceaselessly to do more. This notion may be fueled further by the tenure system at many universities in which it is impressed on tenure-track faculty that one has to be among the very best to receive tenure.

And – to articulate something that often goes unspoken – we also think it is possible that a number of seemingly successful scientists and other professionals bury themselves in their work because the other areas of their lives are not going so well. Perhaps one's married life feels flat or unfulfilling or one can't find a partner to spend time with. Perhaps the pressure of raising children or caring for aging parents is overwhelming or one has difficulty sustaining friendships. Turning our focus to work can distract us from these problems, and provide a sense of control that we lack elsewhere.

This depressing account of the pressures on young scientists may sound familiar to you. Indeed, when visiting academic departments or scientific conferences one does not gain the impression that scientists are particularly joyful. In his book *Don't Be Such a Scientist*, former marine biologist-turned science communication expert Randy Olson (2009) writes that many scientists struggle to communicate both the outcomes of their research and their passion for doing science. As a result they come across as dull or disinterested. Olson provocatively argues that scientists tend to live exclusively in their heads, rarely communicating from other parts of their bodies, such as the guts or heart:

> The doing of science is the objective part. It's what scientists are most comfortable with. A scientist can sit in his or her laboratory all day long, talking to the microscopes and centrifuges, and they will never talk back. I have heard scientist friends of mine over the years rave about how much they enjoy field and laboratory research for exactly this reason – it's all so rational, so logical, so objective, and ... alas, so nonhuman – a chance to get out in the field, away from people. No politics, no bureaucracy, no administrative duties, just pure rationality.
>
> *(Olson, 2009, p. 31)*

Similarly, one of the colleagues we interviewed for this book noted that all kinds of academics – from novice graduate students to accomplished, tenured professors – can "get caught in the performance trap." This trap leads one to believe that we are "defined by the recognition received in accordance with the academic lifestyle – the number of papers published, invited lectures given, research grants won, and awards received. But what happens when the music stops and the dance is over? Where does one then derive his/her sense of value or worth?"

This is exactly the question we are interested in. How does one find joy in and through one's work, but without sacrificing a sense of being a "whole person," or falling victim to the "performance trap?"

JOY AND SUCCESS

The earlier portrait of the academic scientist's life raises the question: Can we experience joy in our laboratories, classrooms, and offices? Can we communicate and share that joy with others? What does it mean to be successful *and* joyful? To be even more philosophical, does *joy* matter in our work? If so, why?

We answer these questions by suggesting that the road toward success is much easier to travel if it is propelled by joy and, in fact, joy is often a sign of success. We think of joy as related to psychologist Mihaly Csikszentmihalyi's concept of "flow," a state of deep contentment one finds when engaged in work or activities that make time seem to disappear (Csikszentmihalyi, 2008). Many scientists may find that they are able to find their "flow," as Olson notes when he describes scientist friends who "rave" about their research. But they struggle to find the same flow in collaborations, or in their home or personal lives. Others struggle with the inverse – they are happy socially and have rich personal lives but struggle to be "productive" at work as defined by traditional metrics, such as publications. The role models we have found most inspiring have found a way to practice joy and be successful in multiple areas of their lives; to experience flow in and outside of the office. Perhaps they are not always happy or "balanced." In fact, all of us get out of whack at times and need to make corrections. But we believe that the practice of joy – and the sharing of that joy with others – is possible across the long arc of one's career. The failure to thrive in this endeavor is one area where we believe scientists may struggle, leading to feelings of frustration or isolation.

We also believe that joy is not the *result* of success. We want you to have achievements you are proud of, and we certainly understand that not all aspects of our work lives can be filled with joy or in flow – some committee assignments come to mind! Yet, one aspect of the performance trap is that we do not often feel joy as a result of our academic accomplishments. We both have seen colleagues who have achieved the next big goal – getting tenure, a promotion, a grant, a recognition – but without being able to truly enjoy it. In fact, this was especially true for Jen, who actually broke out into tears when she heard she had received tenure, because it felt so anti-climatic, and also because she realized she no longer wanted to stay at the university that had awarded her tenure and promotion – a valuable, but painful, realization that we can be successful, but that does not guarantee joy.

Rather than seeking joy in whatever the next success is, we find that joy has to come first, and is actually the main ingredient in building a successful life. The quote at the start of this chapter suggests that finding joy is in fact the definition of success. Success is not our end goal in this model; it is the byproduct of creating more joy in our work lives. For us, "joy" corresponds to a feeling of doing the right thing while thoroughly enjoying it; it is the feeling of being in the right place, having meaningful relationships and making an impact, and being able to declare some positive control over how our lives shape up and shape others. Viewed in this way, finding joy *is* success. Happily, finding joy will often also lead you to be "successful" in the traditional sense of the word; you will find it easier to work with others, solve conflicts, and make choices in your every day life.

We know from experience there are real problems and pressures on academics today, and we don't mean to belittle academics' very real struggles with institutional and systemic problems. Ideally, academics would organize themselves in ways that would allow them to modify policies and procedures that aren't working well for most of us. American universities are undergoing a tremendous period of transition in which job security, academic freedom, and faculty governance and autonomy are far from guaranteed (Gerber, 2014).

Given both internal and external pressures, then, is it any wonder that those of us who stay in science are confused about what finding joy in our lives means?

SOME PERSONAL QUESTIONS

In light of the challenges sketched here, we pose the following question to the reader: How are *you* doing?

This book is meant as a personal guide for developing a fulfilling and joyful career as a professional. It can serve that purpose only when used in a personal way; this book should not be seen as an academic treatise on the academic career. We ask the reader to answer the questions that follow before proceeding. This should not take much time. In fact, we recommend that you do this with little forethought; often our gut feelings and first reactions are more accurate than our well thought-out responses. But we do encourage you to *write down* your answers; it is otherwise too easy to skip over them or to get caught up in familiar internal chatter. In our experience, writing things down helps us to reveal our own thought processes to ourselves, to achieve specificity and clarity on our goals and desires, to question preconceived notions or misunderstandings, and to commit to new paths of action.

It is easy to over-analyze and over-rationalize, so we suggest that you take out a piece of paper to record your thoughts, write by hand (which might help you get out of your left brain), and give your intuitive reactions to the following questions:

- Are you fully and freely expressing yourself? What does this expression look like? How is it received by others? Or do you feel silenced sometimes? Do you find it difficult to say what you really think or feel?
- Do you think of the many parts of your life as being in balance, or in harmony? In what way? What would those closest to you say?
- Do you have personal or professional practices (e.g., carving out writing time, seeking feedback from mentors, meditation) that help you with your internal balance?
- Do you have any dreams in your personal or professional life you would like to fulfill? Or do you struggle to articulate a vision for your life, to explain where you would like to see yourself in five or ten years?

Perhaps all of the questions resonated with you; perhaps you felt a clear response to only one or two of them. That is all fine; there are no right or wrong answers to these questions. This is not a quiz or a test.

Perhaps there is a vague sense of unease with some of these questions.

The main point of this exercise is to get clarity on how you are doing. Perhaps you feel great or are doing great, but there is a good chance that there are areas where you feel some changes would benefit you. Perhaps there is a vague sense of unease with some of these questions. Think of these as pointers – messages from within – that are telling you that you might benefit from doing some things differently. This book serves as a guide to making changes in your life as a scientist that might lead to a more harmonious, less harried life, which we believe will also improve your effectiveness as a scientist.

ABOUT THIS BOOK

We refer to the audience of this book as "scientists" but really we have written this book for scientists, social scientists, and engineers who seek to have effective careers that are joyful and harmonious, and who hope to bring harmony to their daily lives in favor of a more peaceful way of living. We also have thought of our audience primarily as academics, because that is who we are, but we believe it's possible that those outside of the university might also find this text useful. One of the points that we make in this book is that our character, personality, and outlook in life determine not only who we are in our

The authors enjoying commencement.

personal lives but our character and personality also influence our effectiveness as professionals. It really is a package deal; who we are determines how we lead our life, and this includes both our professional and personal lives.

You may be wondering at this point who *we* are, why we feel entitled to talk to you about joy, and why we wrote this book. As you move through these pages you will read more about us and our experiences – and about the experiences of other professionals we have interviewed. In the following, we provide a little bit of information about each of us.

Roel holds degrees in theoretical physics, atmospheric science, and geophysics. His research is diverse, but he gives most attention to imaging techniques, often in close collaboration with industry. He has published more than 260 papers, and this is his third book, his second on professional development for young researchers. Roel is also restless.

This explains his move from Europe to the United States, his frequent change of research focus, and the other activities in his life: between 2000 and 2014 he was a firefighter in addition to being a professor. But Roel also has an innate drive to find *meaning*. This resulted in his work on professional development that includes the courses "The Art of Science" and "Research Ethics" at the Colorado School of Mines. Combining all these activities has been a stretch at times. He considers himself a blessed person; he received a great education and has held good jobs, he was raised by loving parents, he has a loving wife who lives with his quirks, and he has three children. His students and children often think a crazy streak runs through Roel, but this is his way to compensate for a world that is often overly serious.

Jen, on the other hand, is a social scientist and humanities scholar who worked for many years with scientists and engineers at a small engineering college, where she gained tenure, and then made a cross-country move to become a tenured professor in a graduate program located in the city where she grew up. Her classes in policy and communication often take students far outside their comfort zones, because they are invited to use improvisational games, alternate presentation styles, and varied types of writing to think about research and communication in new ways. Jen's academic work is primarily in the field of environmental communication, and much of her research examines the rhetoric of energy industries in light of environmental controversy. Jen is also a wife and mother of two, has a lot of hobbies and interests outside of work, and, at various times, has struggled to find harmony in her own life. Her desire to step out of the "stressed" life and into a more peaceful, joyful way of living has put her on a "seeker's path" where she devours all kinds of readings and teachings about joy and success.

As you can see, we are quite different in our backgrounds and professional activities. Yet there is something that unites us; we both felt at some moment in our careers that we did not fit into a traditional "professor" or "scientist" mold; we felt a certain unease with the shape our professional lives had taken. For each of us that has meant making fairly drastic changes that included, changing universities; emigration; innovating personal development programs at our universities; questioning the "way things are done"; and rethinking *how to be* with our coworkers, families, and friends. Both of us feel that these changes have enriched us. Out of these shared values grew the common wish to help young scientists develop effective and harmonious professional and personal lifestyles. We know from personal experience that making changes is not always easy, especially when the purpose of that change is not yet well defined. Over time both of us have stalled, struggled, and stumbled, and we still do so at times. So please don't assume that we have figured it all out; developing a harmonious life is still a challenge that requires adjustment and practice for us, too. But our hope is that our experiences and seeking will help you to seek your own inner harmony, and to pass on that commitment to harmony to those you mentor.

Making changes in our life by just reading a "how-to" book is not realistic. In general, it is one thing to get insight into a condition, but making behavioral changes is much harder. Making such changes

Such changes can only be genuine and sustainable when they are congruent with your personality.

requires reflection, experimentation, and persistence. Change can often be slow, requiring a commitment to new habits over time. Furthermore, such changes can only be genuine and sustainable when they are congruent with our personality. Indeed, one of the points that we want to make is that our professional behavior is, just like our personal behavior, grounded in our personality and character. It is for this reason that the chapters of this book are organized around the following character traits or habits.

In Chapter 1, we introduce the idea of harmony. *Harmony* means that the many aspects of our selves are working together in ways that bring us a sense of overall well-being. We use the term "harmony" instead of "balance" because we believe that "balance" suggests one must excel in all areas of our lives equally, all at the same time, or else fall off a pedestal and break. Instead, we think it is more helpful to think about building harmonious relationships among our many values and commitments. Sometimes we will prioritize our work commitments; at other times we will prioritize our family commitments, or our health. But if any one area of our lives is neglected for too long, we risk discord in all areas of our lives.

The focus of Chapter 2 is courage. *Courage* is the ability and willingness to move forward even when the task at hand is daunting or scary. Courage is not naïvely glossing over or ignoring what needs to be done; instead it involves facing the task at hand. In this book we describe the courage to investigate who we are and why we do things the way we do them, as well as how to overcome feelings of being stuck, blocked, or stalled. Courage is essential for our personal and intellectual independence, and is thus essential in science.

Chapter 3 examines what it means to have vision. *Vision* is the ability to imagine what it is we want for our work and personal lives, to articulate that desire, and then to design a plan or philosophy that allows us to move toward putting those desires into practice. Figuring out what it is we want is not as easy as it sounds. But practicing developing vision encourages us to figure out who we are and how to stay true to ourselves.

Curiosity is the focus of Chapter 4. *Curiosity* involves having a genuine interest in the unknown. Obviously, curiosity is a main driver of research. But the object of curiosity is not restricted to research. Being curious about other people and one's self is of great value in research and life, and for those working in academia, being curious about students is a prerequisite for good teaching.

In Chapter 5, we focus on the practice of listening. *Listening* is the ability to take in information, but also is about giving ourselves over to our relationship with others. Most of us in academe have been trained to speak, to assert our viewpoints, and to defend our positions. Listening requires the opposite – the ability to be infused with the expression of the other. The issue of listening also raises the question: What, and who, do we listen to? In science, do we listen to common knowledge, or to our own creativity? Do we listen to the outside pressures, or are we attuned to our own goals and ideals? And do we listen to those around us who are willing to help and guide us? Do we listen to our students?

Compassion is the focus of Chapter 6. *Compassion* literally means "with passion." For us as scientists and engineers it not only refers to the passion that we have for the content of our work but it also refers to the way we do our work. Compassion presumes we know that it is "not all about us"; it involves humility and an appreciation that we are an integral part of a network with others that includes our colleagues, and those that we teach or advise. But compassion also influences the way we do and teach science and whether we view the scientific endeavor as a competition or a collaboration.

In Chapter 7, we focus on the last trait of successful, joyful scientists, social scientists, and engineers: integrity. *Integrity* implies that all the parts of us – all the elements of our lives – work together. Integrity means that although we have different roles in different aspects of our lives, we are actually an *integrated being*, the same person who thinks, speaks, and acts at home as at work. It means that we have integrated our personality with the different aspects of our life. Integrity involves a steady focus on being honest with ourselves and others, even when that means being vulnerable, and on staying the course to do what needs to be done.

Addressing these traits and practices in our professional and personal life involves a personal commitment and active participation. As we stated earlier, this book should not be read as an academic treatise on the topic of career development or a how-to manual for addressing issues such as scheduling or promotion. In fact, thinking about these topics in a purely academic fashion can be counterproductive because the cloak of rationality can hide the feelings and intuition that are powerful drivers for how we think, feel, and behave. Instead, this book should be seen as an interactive text. For this reason, each chapter contains exercises. To get the most out of this book, we encourage you to do the exercises consistently – it can be helpful to have a notebook or journal that you write in as you read this book. We also believe strongly in the power of stories, or narratives, in making sense of life changes that we are trying to encourage. Therefore, in order to connect the material in this book with the lives of real scientists, we present in each chapter excerpts from interviews with scientists, social scientists, or engineers who have experience with the concepts we are examining.

I Harmony

> You will never find your perfect life "balance" on the path [of life] for the same reason you will never find a unicorn on the path – because these things don't exist. Forget unicorns and balance. If you were perfectly balanced, you'd never have to take anyone's hand to steady yourself, and that would be a tragedy.
>
> –Glennon Melton Doyle (2014)

THE LURE OF "BALANCE"

Many professionals, including scientists, find themselves wanting to "balance" professional and personal life. The pressure to excel, or be the best, can easily degenerate into a feeling that one can never work enough. The work atmosphere at different universities and other research organizations varies greatly, but there is frequently pressure to spend exceedingly long hours in the office and continue working at home. There is nothing wrong with working hard, but we also find that the temptation to work excessive hours – at the expense of other parts of our lives – is particularly strong for academics, for whom there are often no set work hours, and "deliverables" are never-ending. In the long run, however, such a lifestyle may come at the expense of our relationships, our peace of mind, and our health. Let's not fool ourselves: both life experience and research suggest that overwork actually results in diminishing returns in terms of work productivity as well (Robinson S., 2012).

The "balance" between professional and personal life is particularly difficult for early-career scientists trying to prove themselves on the tenure track (Mason, Wolfinger, & Goulden, 2013). Just when learning curves and pressures are at their highest, these highly educated people may also be considering starting families, creating a perfect storm of demands and expectations. These years may be especially difficult for female scientists who often have the physical

A lifestyle of excessive work may come at the expense of our relationships.

experience of bearing and nursing children and, depending on their circumstances and relationships, may also find themselves responsible for a larger share of childcare and household duties. Scholars are careful to point out that our current academic system in the United States supports neither new mothers nor new fathers particularly well (Mason, Wolfginger, & Goulden, 2013) and there is the perception, if not the reality, that being on the "mommy track" is harmful to one's career. This disproportionately affects women and their choices to remain in academic positions as well as their ability or willingness to occupy advanced positions (such as achieving the rank of full professor or occupying leading administrative roles).

The question of balance, or lack thereof, also applies to research: Are we adventurous in our work or do we want to play it safe? Some research may be very original but have a high risk of never leading to anything useful, while other research could have a more or less predictable outcome, yet allow one to contribute to ongoing academic "conversations" only in modest ways. There thus exists a tension between doing very innovative research, with a relatively high likelihood of failure, and taking a less exciting and more certain approach. This tension plays out not only at the level of the individual scientist; federal and private funding agencies also struggle with this balance between innovation and predictability of success. How easy it is to be out of balance in this tension: as a scientist one might only choose safe, but dull, topics of research, or one could focus exclusively on "pie in the sky research" that never goes anywhere.

There are other ways in which the scientific life could be said to be out of balance. Have you ever looked around you at a scientific conference or other gatherings of scientists, noticing what kinds of people are represented there and also who is *not*? Chances are high that most of those around you were male, and perhaps white males were in the majority. Despite many efforts, and a slow shift toward diversity over time, the scientific community in the United States is still not diverse; 2008 statistics show that whites still made up nearly 80 percent of those receiving doctorates in the sciences, engineering, and health (Milan & Hoffer, 2012). The statistical picture for American women is more complicated: women are well represented in some science and engineering disciplines, such as the biological sciences, but are poorly represented in others, such as physics, and in 2008 they earned 44 percent of US PhDs awarded in science and engineering (Rosser & Taylor, 2009).

While women in the United States are getting advanced degrees in science and engineering fields, that number drops significantly if we look at who continues into science and engineering careers, and numbers globally also show significant gender disparities for women, in both developed and developing countries (Women in Global Science and Technology, 2012). Why is it that women and minority groups are underrepresented? There are historical and social reasons for this, but decades of well-intended initiatives to diversify the scientific community have led to uneven or incremental improvements. As a result, the composition of the scientific community is out of balance. This leads to inequities for women and people of color, but research shows this lack of diversity might also not be good for white males or for science, either: a lack of diversity can lead to problematic practices such as "convenience sampling" and stifled innovation (Medin & Lee, 2012). Some researchers have

The academic life can leave one feeling out of balance, out of sorts,
or decentered.

even suggested that ethnically diverse collaborations can lead to publication in higher-impact journals and increased citation rates (Freeman & Huang, 2014).

We have described several areas in the life of researchers in which a need exists to find some sense of "balance," primarily because the academic life can leave one feeling *out of balance*, out of sorts, or decentered. We may know *something* is wrong but can't put our finger on how to fix it. Or we might be getting signals from others – our department chair, our spouse or partner, our kids, or even our doctors – that we are out of sorts, even if we have been slow to realize it ourselves.

But what does *balance* really mean? Is there a more fruitful way to address our feelings of inadequacy or decenteredness?

HARMONY INSTEAD

Scientists might be tempted to think of balance in technical terms, as an equilibrium between counteracting forces. This view of balance presumes that these forces are conflicting, and that one may at best expect a tug of war between them that results in equilibrium. This view of balance leads to a life wherein it is most important not to "rock the boat," because the imbalance caused by change could upset the equilibrium that we have attained in our life. Maintaining the status quo as the highest priority thus can be the outcome of viewing balance in this way, even if the status quo leaves nobody feeling particularly joyful or fulfilled.

Thinking about balance conjures an image of a vaudeville performer, standing on one leg, precariously perched on a pedestal, all the while juggling many balls in the air. We scientists and academics are that performer, the pedestal signifies "success," and the balls in the air are symbolic of the different parts of our lives. Have you ever had the experience of getting back from a vacation (where you found it difficult not to check your work email or read a report) only to find the relaxed feeling you did have dissipate once you saw how far behind you are when you returned? Or maybe you are a young father who was up all night with a sick toddler (frustrating in and of itself, but also because you know the next day is a busy one at work) who then struggles to feel coherent or productive in meetings and class the next day? In both instances, it is easy to work not hard enough, to not enjoy your vacation enough, to feel you are neither a good father, nor a good employee, nor a human who can really relax. You feel like you're about to drop a few of the balls you are juggling at best, and maybe about to fall off the pedestal altogether at worst.

Another way the balance metaphor proves unproductive is that it suggests we must be "achieving" in every part of our lives. Women professionals are often asked the questions, "How do you do it all?" or "Do you think you can have it 'all'?" Having "it all" means, can you be successful professionally and also be a good mother/partner/house-keeper (or whatever other roles we might conjure)? "Balance" in this worldview means that we are somehow successfully performing in each area of our lives without major conflict or disaster. We're keeping all the balls in the air and staying beautifully perched on the pedestal of success.

The image of a vaudeville performer, keeping all the balls in the air and staying perched on the pedestal of success.

But what if we didn't see balance as a result of constant achievement? What if it wasn't about keeping all the balls in the air at the same time? What if, instead, we thought in terms of "harmony?" In music, harmony refers to the collaboration between different voices to create something greater. The chords in music are more than a collection of individual tones – the tones work together to create a sound greater than themselves. What if we could do the same with our professional and personal life? What if our personal life were to refresh and renew our creativity as scientists, while the fulfillment of our professional life would enrich the personal life that we lead? This view of balance doesn't requires *striving*, a sense that if we could just get our colleagues to take more responsibility, or get our schedule to work better, or reach a particular professional benchmark, we would *achieve* balance. It doesn't require perfection, either, or the sense that we can never stop juggling.

Instead, ours is a view that aims at internal harmony, a state of being and peace that allows us to make decisions in our professional *and* personal lives that serve our whole selves over time and eventually lead to external harmony (Palmer, 2004). Our belief is that a sense of overall well-being cannot help but result. Life is going to have conflicts. There is going to be friction. But developing a joy-centered philosophy for our work and home lives allows us to develop resilience and centeredness when friction does occur. To quote Robert Fulghum, "Peace is not the absence of conflict, but the ability to cope with it" (quoted in Melton, 2014). When we strengthen our sense of *inner* equilibrium, we are much more likely to experience an external equilibrium as well. We do not claim that such harmony is easy to achieve, or that we ourselves are experts at this all of the time. But the practices we describe in this book have worked for us, over time, and also are evident in the role models we most admire.

HARMONY AS GATEWAY TO GUSTO

So, as we noted earlier, "balance" is incredibly hard to maintain, and as a metaphor it probably adds more to stress to our lives than it alleviates. The fact is that life throws plenty of curveballs at us. Adding a layer of shame about getting out of balance on top of whatever difficult thing we are experiencing doesn't help anyone. And frequently, experiencing something difficult gives us important information about areas of our life where we may need to change. It can teach us valuable lessons about ourselves and what is important. One of our colleagues told us this story, about a time when he was in graduate school as a young social scientist who also had a family to care for:

> I was in graduate school, trying to finish my dissertation and find
> fellowships that would help support my family. We had an infant

and a four year old, and my wife was incredibly sick. I had four fellowship deadlines in the space of about ten days, and I just couldn't hold it together. I remember a mental buzz – not the good kind – that had me running from task to task without the ability to focus or really be present. I knew that the immediate biological and emotional needs of my family were most important, and yet I knew too that I just had to make space to finish my grant proposals. I had no room in my life to care for myself, and I just felt like collapsing. There was no wiggle room to ask for extended deadlines, and no way to hire out all of the childcare that was needed.

I remember sitting on the floor in front of my sick wife while she nursed our infant, and just sobbing, saying that I just didn't know how to do it. It was a real crisis, and I survived it with some luck and the love of my wife and some friends and family. It made me aware of how real life can pull you out of balance – that sometimes we need to be pulled out of balance so we know what our priorities are.

This experience resonated with both Jen and Roel – we can each remember similar experiences from when our own children were young and we felt the pressures to be in many places at once, and also the sense that we were doing nothing well.

In her book *Overwhelmed: Work, Love, and Play When No One Has the Time*, Brigid Schulte talks about the concept of "contaminated time" (Schulte, 2014). One of the problems with thinking about work-life balance and time management is that those metaphors and practices do not really address the fact that many of us feel we must be in multiple places at once. We work while we are at home, and when we are at work, we feel as if we are neglecting our home duties, or self-care. We may be "balancing" our work and home lives and health, but we don't have a sense of well-being or joy as a result. What this means is that the hours we spend on both our work and our personal lives feel "contaminated," and ultimately we get little satisfaction or joy from doing either. We are unable to be fully present in either space, as our colleague described. Schulte believes that the answer is to first reduce what we *think* we can accomplish each day (make our expectations for ourselves more realistic) and then practice being where we are, that is, being present at work when we are at work, and being home when we are at home (Schulte, 2014).

One of the reasons contaminated time and aiming for balance are so dissatisfying, in our view, is that they also keep us from experiencing life with gusto. As writer Marta Goertzen puts it, "I need and crave variety. I crave times of intense concentration where I get lost in a project. I long for experiences that push my limits of understanding and push me into discovery mode and force change" (Goertzen, 2014). In other words, intense disequilibrium can bring us opportunities for the greatest personal and professional growth; this is the concept of "flow" we talked about in the Introduction (Csikszentmihalyi, 2008). The physicist Ilya Prigonine makes the point that in physics, chaos and disequilibrium are essential for establishing a new order (Prigogine & Stengers, 1984). There may also be times in our professional lives where disequilibrium leads us into periods of great productivity and pleasure, where we are our most focused and creative at work. And disequilibrium can also help us with time management. As one of our colleagues put it, "When I'm pretty busy I tend to triage things and recognize when something is 'good enough.' This is also when I feel balanced, I think, because I'm more likely to say to myself, 'I only have 30 minutes to work out so I'm just going to do what I can.'" In other words, we can be quite busy, but if we've adjusted our expectations, and remain present where we are, a feeling of intense joy can result.

Goertzen also helps us to think about how we might move from an "equilibrium" model of balance to a "harmony" model of balance by actively choosing to examine the habits we have in place. She writes,

What I do think is that you have to examine your activities and goals on a regular basis. Are you out of balance because you are addicted to the craziness it brings and can't imagine how you will function without it? Or are you out of balance because you have a specific goal in mind and know that short term struggle will bring long term benefits?

(Goertzen, 2014)

For Goertzen, the latter is productive, the former is not. Goertzen's view of balance is useful because it frees us up from feeling guilt or shame for those times when certain areas of our lives are receiving more attention than others. For working parents or those of us with aging parents or health challenges, this can be especially useful: you know there will be times when work simply gets a bit more of your attention so that you can reach a goal. There may be times when your family and friends don't get as much attention as you would like. But Goertzen also wants us not to make the "craziness" a habit simply because appearing "busy" makes us feel worthwhile, or because we think it communicates to others that we are "busy" or productive.

In other words, we all experience times of difficulty, stress, intense focus, or myopia. But those "crazy" times should pass. They should ebb and flow. You must be willing to evaluate your commitments. It can be easy for the stresses of work to always take front seat, and for the demands of our families, bodies, or non-work passions to be ignored.

A key element of maintaining harmony is to question your inner beliefs about what it means to be "busy," "productive," or "successful." Our experience suggests that academics in particular are often vulnerable to internalizing external metrics of success. In lots of ways, our university reward system is set up exactly that way – our success depends on our working hard but also on how others value that work, which is often determined through processes of many-layered external reviews. As a result, it can be tempting to internalize and then perpetuate, rather than challenge, the pressures placed on you by others. It can be tempting to internalize what seem to be the primary values of the academy.

We want you to ask yourself this question: To what extent do *you* buy into the view that science is a competitive activity, taking place in an environment full of stress? If nobody bought into the belief that science is a tough, competitive activity, then the scientific culture would be very different. This means that the pressures do not come exclusively from outside of us (though they certainly do exist). There must also be some part of us that internalizes them as well, thus perpetuating and even institutionalizing them.

Exercise: internalizing expectations

Consider the following statements and note whether or not they are true for you:

- I only feel a sense of accomplishment if I have succeeded in a highly competitive environment.
- Life has to be difficult and full of sacrifices if I want to achieve something good.
- I find it hard to feel happy when others succeed, for it makes me question if I am being successful or productive enough.
- Younger scientists should feel the same pressure that I, a more experienced scientist, felt when I was coming up.
- I achieve more and do better work when I feel overworked or pressured.
- I cannot show that I feel joy, enjoyment, or relaxation when around my peers; instead, I must project that I am always busy or overworked.

Each of us has had moments in our career when we have answered "true" for at least some of the statements in the exercise. Clearly, the system could well be set up to perpetuate some of these beliefs about what successful work looks like. But we have also realized that it takes two to tango, and we find we are much happier *and even more successful* when we are not experiencing those statements as true. When we collectively buy into a system, then we collectively create the system and help that system perpetuate itself. All this happens at the expense of the well-being of a generation of young scientists, as well as those around them. Many may leave the practice of science, feeling that the lack of balance between their work and home lives is too great to overcome.

We believe that the answer to the problems caused by a lack of harmony and feeling overwhelmed by contaminated time comes in being honest with yourself about your practices. For those of us who feel particularly busy, stressed, or powerless, this may prove to be a challenge. We often develop habits in the first place because they help us to cope with our struggles or needs. Sometimes they're good habits, as when we carve out time in our schedules each day to write up our research. Other times, habits are not so salutary, as when we find ourselves storming across campus in a constant state of stress and hurry, not stopping to greet students or colleagues, or to enjoy the fresh air. We may be consistently late for meetings, or feel surly about them, because we've overbooked ourselves. This habit serves no one. A good habit, on the other hand, could involve making time for physical exercise so that our minds, hearts, and bodies feel fresh. A bad habit might be pulling out your cell phone to check work emails when at dinner with family or friends. Whether habits are working for you and your loved ones cannot always be answered rationally – you may need to check in with your heart, and with your loved ones, to know if you are on the right path.

HARMONIZING OUR INNER AND OUTER WORLDS

We experience life in the material world, the world of events, with the people around us. We call this collection of external events the "outer world." In addition to the outer world, there is the "inner world" that consists of everything that we experience in our mental realm; this is the world of our thoughts and beliefs. Our senses connect the outer world to our inner world, so we could see them as simple communication channels between the outer world and the inner world. However, our sensory organs don't just send inputs that are then presented to an "awareness"; our sensory organs already pre-filter inputs so that they help perceive patterns that prevent an overload of information, and thus create order. The situation is even more complicated in that our mind dictates to a large extent how we perceive the world. We may shut off our ears to unwanted sound or we may focus our hearing to something we keenly want to listen to. You may know the experience of searching for an object that was lying in plain sight, but you overlooked it because you did not expect it to be where it was. We may literally not see what is in plain sight because of the filtering done by the mind. Likewise, we may not hear, not because we are deaf but because our mind is closed off to the message.

Our perception is colored in deeper ways by our minds as well. Our general mood clearly changes our perception. On days when we are optimistic and full of internal energy, everything we encounter looks beautiful, people seem to be upbeat, and when they are not we simply shrug it off. By contrast, on days when we feel down, the world literally seems dark, and challenges appear to assault us from all sides.

But the world outside us is also shaped by our thinking. This is obviously the case when we undertake specific actions and change something in the world, but it also happens in more subtle ways. Our thinking sets events in motion in unpredictable ways that we often don't even understand. This may happen because we subconsciously do something or communicate something that produces a result that comes forth in our life. For example, Roel was once at a conference, and he noticed he was both grumpy and skeptical and mustered little interest in those around him. After some reflection, he realized that all it would take to change this was to decide to be truly interested in others and in the work being presented. After he became aware of this, he put his intention on being present and engaged with the people around him and what they had to say. This shift of awareness changed everything! He met interesting people, had meaningful conversations, and learned of exciting science. The conversations he had on that day led to new research projects for his students, and invitations for events. Whatever the mechanism between the inner and outer world may be, these worlds are closely intertwined.

In this connection between the inner and outer world, our beliefs are important because they help shape not only our perception of the world around us but also the events that unfold. The way we view the world is shaped by our beliefs. When we go to a conference and are convinced we will encounter creative ideas and inspiring people, we will have a completely different experience than when we go to the meeting expecting that there is nothing new under the sun. Our beliefs are the anchor point in our inner world.

Exercise: map your inner and outer worlds

The earlier discussion shows that our experience is shaped by the combination of external circumstances, our perceptions, and our beliefs; these represent the outer and inner worlds. In this exercise we ask that you draw one map of your outer world and one of your inner world. The map of your outer world is a physical representation of your external commitments and activities. It might contain symbols of or locations for your work, family, friendships, communities, hobbies, and interests.

The map of your inner world, on the other hand, may contain elements such as your dreams, ambitions, fears, beliefs or faith, and principles you want to live by.

Feel free to draw the maps in any way that you like, but we encourage you not to make a list; because of its linear character, a list forces us into linear thinking. Drawing, doodling, using mindmap software, or arranging pieces of paper can all productively lead you to more intuitive or creative forms of thinking.

After having drawn these maps, we invite you to examine them and to consider in what ways the maps of your outer and inner worlds are aligned. Such alignment, or lack thereof, is important because it will help you discover the degree to which your inner and outer worlds are in harmony with each other. A lack of such harmony might help you discover whether there are changes that you would like to make in your professional or personal life. It is useful to write down what you discover. When we just think, our thoughts often are vague and noncommittal. When we write something down, we need to be specific, and the action of translating thoughts into words makes us in practice much more committed.

We suggest that you keep the maps and your written comparison of these maps, because you may find it useful to revisit them in the future, and perhaps change them later as you are changing. This is the beauty of self-reflection: if we muster the courage to face who we are today and who we want to be tomorrow or next year, we can change.

Be prepared, as you think about doing this kind of self-evaluation, to encounter feelings of resistance. You may have some unexamined beliefs that keep you from making choices that allow you to harmonize the competing demands on your time and spirit. Examples of such beliefs might include the following:

(1) I cannot be a successful scientist when I also spend time out of the office.
(2) The more I work, the better the outcome is.
(3) My supervisor is never going to promote me unless I sacrifice everything for my work.
(4) Wherever I look, I see others that are better than I am.

The first belief – that life is mostly lived in the office or lab – is common in academia, and the roots of this belief are often effectively sowed in graduate school where advisors, or fellow students, imprint on new students that one is expected to "always" be in the office. There may be good reasons to set rules that ensure that adequate work hours are made and that there is sufficient common time to be in the office and interact with colleagues. But such rules do not necessarily include that one needs to be in the office every night and weekend.

The second belief, "the more I work, the better the outcome is," is a subtle variation on the first belief because it articulates that more work always leads to better outcomes. In contrast to the first belief, where presence in the office is required, this second belief about overwork fuels the belief that "I can never do/be enough." This belief is not only detrimental to building harmony in our lives, but it may also impact our science negatively because we never take the distance needed for truly new insights.

The belief that promotion will only take place once we have sacrificed everything else is related to the other beliefs, but it contains the additional element that somebody else will only take certain actions, in this case promotion, when I behave in a certain way. What is dangerous about this belief is that is might actually be wrong. This belief is especially relevant for scientists who are on the tenure track. This is indeed a period in the scientific career where scientists are vulnerable, but the notion that one must sacrifice other parts of one's life to get tenure, promotion, recognition, or awards is a particularly pernicious cultural problem in science that must be addressed at both the institutional and professional levels.

One of us, Roel, ran into the fourth belief – that others are always working harder or doing better than we are – when balancing a demanding job in academia with care for three young children. At work there were colleagues that were more productive, and while taking care of the children, Roel met others whom he felt were more devoted to being a parent than he was. "Juggling" the scientific career and raising a young family thus led to a feeling of being inadequate everywhere; there were more productive scientists at work and better parents outside work.

This sense of being inadequate was resolved by a friend who asked the following question: "Does the fact that you *could* do something better mean that you are not doing a good job *now*?" This question changed everything, because it made it clear that you don't have to be the very best to contribute in a meaningful and useful way. Similarly, one of our colleagues shared the advice his postdoctoral mentor gave him about tenure: "Don't work too hard for tenure; you might not get it anyway." This might seem like discouraging advice, but our colleague believes his mentor meant that tenure was not always based on merit or effort, and that we must remember to care for our families, bodies, and hearts as well as our personal lives. Our colleague explained, "If we fall into the trap of truly believing that if we just worked harder, we would achieve the next reward, then we risk ramping up our work life to a momentum that ignores the joy of being a whole person with many dimensions beyond work."

If these beliefs, or variations of these beliefs, look familiar, you may want to re-examine these beliefs, and perhaps replace them with alternatives for a more balanced, and eventually a more productive, life. Such alternatives could include beliefs such as "getting out of the office and moving my body provides the clarity of mind to be a creative scientist" or "in order to be good at something and to contribute in a meaningful way, I don't need to be the very best." Obviously, it takes courage to create such a different mindset, especially in a work environment that buys into the idea that whatever you do, "it never is enough."

Exercise: examine your beliefs

As we described earlier, our beliefs are important in the way we perceive our life, and give direction to how we show up. Take a few moments to consider the following questions in your journal.

- What are your beliefs about "success?"
- How would you describe your skills and abilities at work?
- What do you think of the contribution you make?
- What judgments do you have about your personal life, and your ability to create "uncontaminated time?"
- Can you think of colleagues or students that you judge for working too much, or not enough? Are there ways in which you also apply those beliefs to yourself?
- In what ways might these beliefs be untrue?

2 Courage

We argue that doing science is, in some ways, inherently courageous. Finishing a college degree in science, then going to graduate school, on to a postdoc, and then on the job market may all take great courage, depending on your origins and circumstances. Furthermore, scientists usually are inquisitive people whose job it is to carry out innovation and discovery. For this reason, we think of the scientific community as being, in some ways, very open-minded to change and inquiry. Our personal experience also tells us that many scientists are comfortable challenging the status quo and authority – whether academically or politically – and many speak their minds freely.

This spirit of inquiry is often essential to doing the work of science. To move our work forward, we may have to let go of things we have learned or suspend our tendency toward disbelief and cynicism. A case in point is the development of the theory of special relativity by Einstein. A cornerstone in this theory is the assumption that the speed of light is always the same. At first sight this seems preposterous: How can two observers that move with respect to each other see the same photon move at the same speed? Yet by simply assuming the constancy of the speed of light and logically deriving its consequences, Einstein changed our worldview with the theory of special relativity. He admittedly did not make his assumption out off the blue – there were experimental indications – but it still took great courage to build a new theory on a foundation that seemed to fly in the face of common "knowledge."

On the other hand, we also believe that the scientific community is somewhat conservative, meaning it resists change, especially when the concepts and ideas underlying the change are unfamiliar. Scientists often approach change with skepticism because skepticism allows them to critically evaluate new information. Such resistance is described in great detail in the book *The Structure of Scientific Revolutions* by Thomas Kuhn (1962). He makes the point that progress in science is not one of steady growth and evolution but that science moves forward incrementally, until faced with disruptive changes. These disruptive changes, or scientific revolutions, are interspersed with periods in which there are few fundamentally new insights. At that point the scientific community focuses on implementing the insights gleaned in the past revolution, on making incremental adjustments, or on everyday business. Without indications that a new concept or insight is needed, the community settles into happy contentment that it knows how things work.

It is, in the light of this contentment, not difficult to understand that the renegade scientist who proposes something new may face skepticism and resistance. To a certain extent such skepticism is healthy because it helps the community separate the chaff from the grain. Processes such as peer review make the creation of scientific knowledge over time both possible and reliable. But too much skepticism or resistance to change may be unwarranted or even unhealthy. When skepticism is driven by an egotistical desire to defend one's turf, it usually just stifles innovation. Whatever the nature of the skepticism may be, it takes courage to push forward and to stand up for the ideas and insights we are proposing.

Another historical example illustrates what we mean. In their book *Made to Stick*, Chip and Dan Heath describe the story of medical researchers Barry Marshall and Robin Warren, who discovered in the 1980s that stomach ulcers were caused by bacteria – not by stress or diet, as previously thought (Heath & Heath, 2007). This was an important breakthrough because it meant that these ulcers, which could be uncomfortable and even painful, were also quite curable through the use of antibiotics. Heath and Heath write of the tremendous opposition and even ridicule that Marshall and Warren faced from their medical colleagues when they made their findings public. "Marshall and Warren could not even get their research paper accepted by a medical journal," Heath and Heath write. "When Marshall presented their findings at a professional conference, the scientists snickered" (Heath & Heath, 2007, p. 131).

Eventually, Marshall decided the only path left open to him was to drink a fluid infected with the ulcer-causing bacteria itself, *H. pylori*. Sure enough, he made himself very sick with an ulcer, and then "cured himself with a course of antibiotics and bismuth (the active ingredient in Pepto-Bismol)" (Heath & Heath, 2007, p. 132). Still it took another ten years for their findings to be widely accepted by the medical and scientific community. At last, some 20 years after their discovery, Marshall and Warren were awarded the Nobel Prize for their work.

We are not necessarily advocating that every scientist take the extreme route that Marshall did. Social studies of science are also littered with examples of researchers who took very public risks to make their discoveries known, only to be proven wrong – the "cold fusion" case being one notable example (Collins & Pinch, 2012). On the other hand, most successful academics we know have had some failures along the way to making courageous scientific discoveries; they are often willing to buck pressures that attempt to dictate what scientific or professional lives and work should look like. Risk-taking and failure are frequently a precursor of success. One colleague we interviewed told us the story of how, as a PhD student at Columbia University, he had to completely revamp his research trajectory:

> After more than two years of working on my research problem, it dawned on me that my research was not going to be useful for my field. I even had my doubts if it qualified as a dissertation. Doubts gave way to dread, and the thought crossed my mind more than once that I might not be able to graduate. Why didn't I recognize this early on? I was too naïve and I did not want to admit I wasted more than

two years of research, hoping I could somehow pull a rabbit out of the hat.

During the third year of research, I went home for Christmas vacation to decompress with my family. At the end of this vacation, I decided to completely abandon the previous three years of research and start anew. Within about six months I had completed most of my "new" thesis. I confidently defended the following winter.

Several decades later, I read that a successful scientist is someone who not only asks the most important research questions but also has the courage to recognize and terminate a research effort headed towards a dead end. I think this last characteristic perfectly describes my dissertation experience. I also gained the fierce confidence to travel my own path, not that of others.

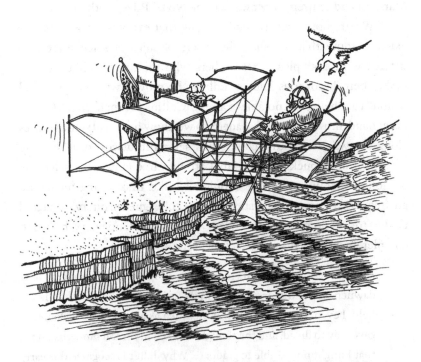

We see many cases of colleagues who courageously take scientific and professional risks.

From our vantage point, then, we see many cases of colleagues who courageously take scientific and professional risks that pay off, both in terms of external reward and also internal satisfaction. These colleagues listen to an internal voice that tells them they are onto something, they reflect on their own responsibilities and desires, and they then press forward despite what critics – or their own internal critics – might say.

THREE WAYS TO THINK ABOUT COURAGE

In this chapter, we invite the reader to think about courage in three ways: courage as "heart," courage as "right action," and courage as "self-reflection." The first comes from the root of the word "courage" which author and scholar Brené Brown tells us is the Latin word *cor*, meaning heart (Brown, 2015). Courage in this sense is a change of orientation that allows one to experience the world and make decisions using both the heart *and* the head. Randy Olson (2009), the former marine biologist and science communication expert whom we referenced in the Introduction, argues that while

Some scientists lose their ability to meaningfully connect and communicate with others by avoiding the lower organs such as the heart and the gut.

most scientists live in the "head," successfully reasoning their way through most challenges, they lose their ability to meaningfully connect and communicate with others by avoiding the "lower organs" such as the heart and the gut, which is where intuition and human connection reside.

The second way we think about courage in this chapter is in terms of self-reflection, being willing to question our own habits, and in particular our own *habits of mind*. We think of this as internal work, the work of looking inward to analyze our perceptions and beliefs. Scholars, and our own experience, tell us that humans are storytelling beings above all else. We make sense of the world through the types of narratives we tell about it. Even those of us who value objectivity must subjectively piece the world together using our perceptions and past experiences to make sense of things. Sometimes those narratives or stories, which seem so real and fixed, need revising so that we can have a hand in designing the lives and careers that we truly want.

The third and final way we use the word "courage" in this chapter is to describe the willingness to *act* on what is right, and what will make us feel more alive with integrity and joy, even in the face of resistance. In the words of Glennon Doyle Melton, courage in this framework looks like "showing up and doing the next right thing" (Melton, 2014). But a word of caution: as we noted in the previous paragraph, we can sometimes be wedded to believing that we are always right. That is not what is meant here by right action. Figuring out what is "right" in this context is not about reasoning ourselves into dominating others through our supposedly superior intellects. The "right" here refers instead to our intuition and to acting from a place of harmony, from a strong sense of self. That may require getting quiet with ourselves before reacting or making decisions. It may require giving up our sense of self-righteousness or need to demonstrate our intelligence to others. This can be tricky because our scientific training often encourages us to flaunt our intelligence or to respond quickly. Showing up and doing the next right thing requires paying attention to our intuition, contemplation, and then acting from a place of harmony, and not simply reacting to circumstances.

COURAGE AS HEART

As scientists, we tend to collectively accept the notion that science is above all driven by rationality and reasoning. After all, isn't science a logical activity that is driven by our rational intelligence? Since science tends to attract people with a strongly developed rational intelligence, it is easy to believe that one only needs rational intelligence for being a good scientist. There is, however, a paradox in science, which is this: although science relies on logic to do its work, science often progresses in ways that have nothing to do with logic (Snieder & Larner, 2009). Instead, the progress of science is often driven by other ways of thinking that include creativity, intuition, passion, free association, serendipity, play, and interaction with others.

We don't mean to belittle logic and reason; we value them greatly. We're also not suggesting you abandon your data, the scientific method, or your analytical viewpoint. Our own work as academics and scientists relies heavily on each. We are suggesting, however, that thinking through professional and personal interactions from the place of the heart offers a gentle corrective to our tendency to be overly positivistic or mechanistic in those interactions. We find that, as scientists, our tendency in our interactions with others is that we want to be "right" above all else, even if that means sacrificing authentic connection with others or with ourselves. Courage allows us to give up our commitment to always being right. Courage helps us sometimes *feel* our way through situations and decision making as a form of reasoning that can be just as valuable as logic.

We also think bringing heart into the practice of science could make us better scientists because it helps us tap into our creativity. Living with heart, with courage, also leads to living with more gusto, which in turn leads to more fulfillment and joy. Increasing our sense of harmony in this way goes hand in hand with better scientific practice and discoveries. Gusto describes your ability to act courageously from a place of great joy and connection with others. It is the opposite of drudgery as a model for one's work life, and it encourages us to see linkages we hadn't thought of before, new collaborations, and even "eureka" moments.

This means that much more than the rational mind is at play when moving science forward. In the larger culture, we often describe the rational mind as the "left brain" and the emotional or creative mind as belonging to the "right brain." Though neither Roel nor Jen are cognitive experts, our understanding is that this is a vast simplification of complex cognitive processes. It can be very difficult to discern between "rational" processes, which may involve "non-rational elements" such as gap-filling, pattern identification based on incomplete data, and narrative sense-making. And "non-rational" processes often have a logic of their own, drawing from cultural and personal experiences that have great significance to us as humans. Our understanding of these processes is in fact new and still evolving. We use the left brain/right brain split here not to describe actual physical or cognitive processes, then, but as a metaphor for how we, and our culture, think about the rational self in distinction from the intuitive self.

Obviously, the right brain is important for many activities other than science. Daniel Goleman's groundbreaking paper on emotional intelligence discusses the importance of the right brain for leadership. For Goleman (2004), emotional intelligence has the following characteristics:

- self-awareness
- self-regulation
- motivation
- empathy
- social skills.

Goleman argues first that emotional intelligence is important for good leadership, and second, paradoxically, in choosing leaders we often select against emotional intelligence. For example, research has shown that leaders with empathy are more effective than those with little empathy, but in selecting leaders we frequently select against empathy because it is perceived as a sign of weakness.

What does this have to do with courage? Using the right brain, or emotional intelligence, is still very much a taboo in the scientific community. Speaking about the "soft stuff" is often frowned upon; instead one has to maintain the façade of being a tough, rational scientist. The very fact that personal and emotional skills are described as "soft" is problematic – these skills often take great strength to practice, and "softness" is also stereotypically gendered language, which we find overly constraining. So it takes courage, not only to use the right brain but also to talk about the issues of the right brain, and to reason not simply from our intellects but to acknowledge that we integrate all sorts of value systems and judgments into our decision-making. In fact, we are often better decision-makers and problem solvers when we acknowledge and nurture these many facets of ourselves. And when we do so we bring our inner world to the outer world, as mentioned in the quote of Laurens van der Post at the start of this chapter.

Exercise: How is your emotional intelligence?

As described earlier, the hallmarks of emotional intelligence are self-awareness, self-regulation, motivation, empathy, and social skills. Describe your strengths and weaknesses in each of these areas. How can you grow in the areas where growth would support you?

COURAGE AS SELF-REFLECTION

Most of us academics have a number of hidden beliefs about what can and cannot be said or done in our professional lives. One example is the belief that, as scientists, our careers must be the most important thing in our lives. We mentioned earlier that we often experience great internal resistance against changing a belief, and this resistance can even be larger when it concerns a collective belief. The most effective defense mechanism against such change is to refrain from talking about it, because the conversation itself might lead to a change in the belief. This leads to an unspoken social contract to not discuss a widely held belief, even when it leads to actions that are just not working for the majority of people – such as a culture of overwork.

If we believe that everyone around us believes that overwork is healthy, for example, we are likely to internalize that belief. It can take tremendous courage to speak from within this spiral of silence and articulate a different value system. One of our colleagues shared the following with us:

> In recent years I have become more tenderhearted and
> compassionate than ever before. Unfortunately, this results
> from personal tragedy in my family. My son, Luke, struggled
> with depression for twelve years. I believe his depression was
> born out of the performance trap – his efforts to "measure up" to
> what he thought was expected of him. He was naturally gifted in
> many areas – academics, athletics, and music. But he was
> focusing too much on pleasing others and not enough on being
> true to himself. It sucked the life out of him. After wrestling
> with his depression for twelve years, he finally decided to end
> his life. Of course, this was a devastating tragedy that became
> a defining moment in my own life. I came face-to-face with the
> truth of the saying that we are "human beings" not "human
> doings." Our value as individuals is innate in who we are and
> does not depend on what we do in life. This is a message that
> I try to convey in advising and teaching students. A student
> should not feel he or she is defined by the grade received on the
> latest exam.

At the onset of Luke's mental health issues I was caught up in my new role managing geophysics research in the corporate world. In retrospect, I think the healthiest and perhaps most lifesaving move on my part would have been to put my career on hold and take a 3- or 6-month leave of absence to get away into a relaxed environment with Luke so he could decompress, reassess, reboot. Perhaps we would all have reassessed our lifestyle issues and made some major changes that would have restored Luke to robust long-term health. Who knows? Anyway the main point is not to get lost in the trees in one's pursuit of his/her career, but to be able to back off and take a look at things from 36,000 feet, perhaps even putting the career on hold for a healthy "time out."

This is a story of true courage. Surviving this tragedy is courageous. Sharing this story with others, rather than hiding one's grief and loss, is courageous. Allowing one's self to become more "tenderhearted and compassionate" as a result is courageous. All of these actions represent a willingness to live and act from the heart, and a willingness to reflect on how we might live differently as a result of our experiences.

Our colleague's story about his son Luke is also a reminder that many of us – even those who seem to be very successful and accomplished (stars in our professions!) – suffer from the unspoken fear that we don't truly "belong" where we are. This may manifest as a sense of unease with our current work situation, or as feelings of discomfort and isolation. Perhaps the feeling of not belonging arises because we really care about things other than the topics we normally discuss with our colleagues, or because we have somehow ventured into a research direction in which we would rather not go. Women and ethnic minorities may also feel they don't belong in the dominant cultures of science and engineering, depending on their work environment.

Exercise: do you belong?

Think about our colleague's story, and about your own experiences in professional and social settings. Then reflect on the following questions in your journal.

- Are there ways in which you feel like you don't belong, or are "pretending" in your professional and/or personal life?
- When do you feel most yourself, most authentic?
- Are there times when you put on a mask, depending on the context?
- Do your colleagues or students struggle with feeling isolated? How do you know?
- How could you create more of a sense of community with, or for, people in your work and home lives?

A colleague told us that one of the main challenges she faced as a graduate student, and that she sees her own students facing now, is the feeling of being inadequate, or like they are "pretending" at being a scientist. Even worse is when students really are pretending to belong as scientists or academics, when they know they would prefer to do something else. She wrote, "Here's the worst ... students going to grad school in a field they don't enjoy, but it's the next step in their master plan they started in high school. They're doing what they're 'supposed to do' and not what they want to do." We would add that this is true not only for graduate students but even for those who may be far along in their careers.

Staying silent about our feelings of isolation usually makes them worse, and can contribute to the sense of feeling trapped or stuck. Starting a conversation on this topic with a trusted colleague might bring the pleasant surprise that you are not alone in this concern, and such a conversation might lead to useful connections that alleviate your isolation, or to a shift in the scope of your activities that revitalizes your work.

In a more extreme form, the feeling of "not belonging" can lead to, or exaggerate, the *imposter syndrome*. Valerie Young writes about

The feeling of "not belonging" can lead to, or exaggerate, the imposter syndrome.

the imposter syndrome as being typified by the sense that one does not really deserve one's position or recognition, and therefore that one is a fraud who is bound to be discovered at any moment (Young, 2011). For example, you might be haunted by a sense that the search committee that selected you made a mistake. Women and minorities may fear that they were appointed because of their minority status, instead of their accomplishments. Students may feel as if their professors are just having pity on them when they perform well, justifying away their good grades or achievements. All of these are instances in which the "story" you tell yourself serves to isolate you from others and worry about your value. You may or may not be aware of this, but we tell ourselves stories most of the time to a degree that one can speak of having an internal dialogue.

Truthfully, the impostor syndrome is common, for all kinds of people. You may suffer from it, but so might your students, colleagues, and mentors. Even those who seem incredibly confident may harbor fears of inadequacy. For this reason it is important to refrain from feeding this syndrome in yourself or in others by paying attention to the language you use to describe your accomplishments or status, and those of others. Often, our colleagues and students need to be affirmed and empowered, provided that such encouragement is genuine. And we can provide the same to ourselves by forming supportive communities that allow discussion of these topics and that ward against isolating thoughts and experiences. University administrators concerned with improving campus "climates" and faculty morale would do well to support initiatives that provide these kinds of opportunities for employees and students.

Finally, courage as self-reflection also requires us to take a hard look at the ethics of our professional choices. Another topic that is not examined enough in science or engineering is whether our research itself is the right thing to do. Yet many of us have such questions at times, questions that are ultimately about the ethics of our work. In fact, it is very useful to reflect on the content of our work because the choice of what we devote our time, energy, and creativity to is one of the most important, far-reaching, and ethical questions in research. But when we gather in our professional groups, we often don't discuss this search for meaning.

While many humanities and social science disciplines have made self-reflection central to conduct of research, we are not aware of many scientific disciplines that similarly promote such introspection, beyond micro-ethical "responsible conduct of research" concerns (Herkert, 2003). Instead, scientists often present research non-reflexively, without taking the time to interrogate how we contribute to the world, and whether our contribution is ethical and meaningful. Instead, as scientists and academics, we tend to glorify our work and have a tendency to overstate the importance of our research to society or to the environment. It takes courage to instead question whether our research is worth pursuing, and to examine not only who benefits from it but also who might be harmed. It takes courage to ask such questions because the answers might imply that we have obligations or responsibilities to our fellow humans that are inconvenient or complex. Such inquiry may require us to change our research and ourselves.

The old adage goes that you shouldn't discuss politics and religion with coworkers. But we would argue that acting as if science and scientists were apolitical, and without spiritual, religious, or philosophical commitments, is problematic. Science and Technology Studies (STS), a field that examines the practice of science as its focus, has persuasively argued for decades that science is inherently political because it acts upon and within a social context. Furthermore, cultural battles such as that over evolution versus intelligent design have encouraged some of us to think of religion or spirituality as dirty words in the scientific community. Similarly, the social conversation about humankind's contribution to climate change has been given a political and ideological twist that dominates its scientific aspects in the eyes of the general public.

Yet there are those among us who are interested in politics or religious and spiritual growth *and* who do rigorous, reasoned

Conversations are only effective when held in an atmosphere of dialogue rather than debate.

scientific work. It can be scary to raise such issues with our colleagues. And sometimes to do so is inappropriate. We may need to build our interpersonal skill set to know when and how to raise such conversations. Sometimes doing so within the safe space of a reading or discussion group is best. Moreover, such conversations are only effective when held in an atmosphere of dialogue rather than debate (Bohm, 1996), which means that, once again, we must be willing to let go of the desire to be "right." In any case, it can take courage to start such a conversation, but the opportunities to expand our understanding, compassion, and connection may make it worth it.

COURAGE AS RIGHT ACTION

Thought or reflection without action is not enough, of course. So courage also refers to the willingness and ability to move forward, even when doing so involves facing resistance. As we noted in the introduction, we do *not* believe that courage involves plowing forward while mindlessly ignoring what the "still small voice" inside of us says. Being courageous also does not mean being foolhardy or naïve. Instead, we believe courage involves getting quiet, listening to our intuition, taking others into account, and then figuring out the next right thing. We can only fully commit to a course of action if we've figured out why it is right to begin with. An essential element of courage thus consists of fully giving ourselves to what we have set out to do. That does not mean we are oblivious to the complications we may run in to, but we choose to just deal with such complications rather than to let ourselves be paralyzed by them.

There are many ways for us to practice this kind of courage as scientists. We might propose a new theory that is likely to generate resistance from colleagues. We might experiment with new teaching techniques in the classroom that push us out of our comfort zone, that might fail and make us look foolish, or that are frowned upon by others. We might act from the heart by communicating to our students that we really care about them. We might stand up for causes on our college campuses that we know will make life better for others.

Perhaps you are an accomplished professor who believes in the importance of diversity, and so you will argue for the creation of a daycare on your campus to improve the lives of professors/parents who are your colleagues. Or you will argue for more inclusive hiring and recruitment practices, or speak out against racist, homophobic, or misogynistic language in classes or on the workfloor. Or, like Roel, you will start a campus reading group that encourages engineering and science faculty, students, and administrators to come together and discuss important issues such as social justice, cognitive science and learning styles, and work-life balance. Instead of reacting in each of these cases with defensiveness or righteousness, we quiet ourselves, figure out what values we wish to stand for, and then show up.

> ### Exercise: where do I not show up?
>
> We explained earlier that it may take courage to show up. Is there an area, or perhaps several areas, in your life where you don't show up? In your journal, write down what this area is, and describe in what ways you feel you don't show up. What would you have to change to be more courageous or present? Pick one of the areas in your life where you don't show up but would like to, and write down three steps you could take to change.

Courage as right action also invites us to be vulnerable, because some-times "showing up" makes us feel uncomfortably visible, or forces us to deal with conflicting views. There are many times when it is much easier to not show up, to keep our heads down and our office doors closed and to just do our "work." Showing up and doing the "next right thing," however, involves risk-taking and the possibility of failure, which can be frightening, especially when our showing up is frowned upon by colleagues. We might come face to face with our own privi-leges or roles in contributing to inequities or injustices. We might have to apologize or correct our mistakes. Others might see our weak-nesses. But we believe strongly that not showing up means you are depriving yourself of possible connection with others simply because you fear conflict or responsibility. Life without courage is life without gusto and real connection. And connection, when nurtured and sustained over time, is a significant source of joy.

Finally, courage is a habit or character trait that we each possess in different degrees. It is a property of our personality, but it is also one that can be built and nurtured over time. So we can and should share courage and promote it in others. This is what we mean when we talk about *encouragement* – helping another to act from the heart, or to "take heart." As scientists we may be in a role where we supervise others, whether they are students, postdoctoral fellows, or other researchers. In addition to such a supervisory role, we also work with colleagues. Fostering the growth of courage in others is intensely rewarding; creating opportunities for others to live their fullest and brightest lives is one of the great privileges of being an academic.

Furthermore, encouragement offers an important counterbalance to the culture of "never enough." Roel received a degree in theoretical physics in a five-year program. Later, in graduate school at another university, Roel's advisor at one point put his arm on Roel's shoulder and said, "I know you will do a wonderful job with this project." Even though this may seem to be a perfectly natural thing to say, at the time it felt like a startling statement to hear, not just because of the encouragement that it conveyed but because in Roel's academic education it was the first time ever that somebody said something like this. Our young people often are uncertain and insecure, and offering encouragement and affirming their potential are among the greatest gifts we can give them.

3   Vision

Your soul knows the difference between a calling and a career.

Clarissa Pinkola Estés (2011)

VISION AS PRACTICE

Reading a chapter titled "Vision," you are probably preparing yourself either (1) for instrumental advice on how to set out goals and meet them, or (2) a more "woo-woo" approach that's going to ask you to imagine yourself sailing on a yacht (or whatever it is that most appeals to your inner fantasy life).

Actually, we think working on a vision for your life is both instrumental and creative; it involves both serious soul-searching or reflection *and* playful imagination. Developing your "vision" muscles is what will allow you to figure out what it is you most want to spend your time doing at work and at home. Having vision contributes to your sense of harmony across all parts of your life. It helps you figure out what "success" looks like for you, and it will aid you in articulating which opportunities you emphatically wish to embrace and which you need to turn down.

We've all heard stories of professionals who seem to have a laser-like commitment to a particular goal, and regardless of what obstacles life throws at them, they keep at that goal until they achieve "success." This type of story might be particularly appealing or familiar to scientists: the stereotype of the lone researcher, toiling away in the lab against all odds, is still fairly prevalent in Western culture. For most of us, at some point in our careers, we seek out advice that will help us hit that "success" mark – a decision guide that will help us figure out how to navigate the many duties, responsibilities, and pressures on our time.

But many of us eventually find that calibrating our notions of "success" to external barometers, whatever they might be, leave us

Imagine whatever it is that most appeals to your inner fantasy life.

feeling adrift and powerless to determine what it is we want for our lives. Perhaps you are a graduate student trying to figure out which schools, projects, or advisors are best for you. Perhaps you are an early career academic who is having trouble focusing your attention – you feel pulled in all directions, or tenure and promotion expectations are not quite clear, and you receive lots of conflicting advice. Or perhaps you just received tenure, and you are wondering, now what?

Jen, for example, remembers being on the tenure-track early on, and receiving a lot of advice about what to do with her office door, because accessibility – being seen as being available and collegial, or conversely as busy or preoccupied, signaled "success." Some colleagues advised her to "always keep her office door open," so that she could chat with colleagues or students passing by, foster collaborations, and be seen as collegial, a team player. Other colleagues recommended she keep her office door closed as much as possible. That would communicate that she was a serious researcher who was able to carve out writing time and effectively say "no" to superfluous conversations and requests.

So which is it? Do we choose to keep our metaphorical office door open or closed? How does a young professional figure out the answer to this question, and to what to do with his or her time? Which advice is right?

The answer is this: it depends. It depends, specifically, on what vision you have for your professional life. Every decision you make – whether to work today with your door open or closed, whether to take this committee assignment or that, whether to attend a conference or your friend's birthday party – should be guided by this vision. That is what this chapter is about: developing this vision and then figuring out how to persist with it.

After we had finished conducting our online interviews with colleagues in preparation for writing this book, we noted with some surprise that a number of professionals we talked to noted a specific instance in their lives in which a *vision* for how to live and work became crystal clear to them. Frequently, these instances were a result of crisis, as with our colleague who found himself with young children, a sick wife, and a number of proposals due, all at the same time. This crisis point allowed him to articulate his priorities, a lesson that stuck with him many years after the crisis had passed. Sometimes, as with our colleague who lost his son to suicide, these moments of clarity came as a result of tragedy. Tragedy can also sharpen our ability to prioritize our values and begin to build a life with more heart, more courage. Another colleague shared this story with us:

> Between undergrad and grad school I worked a "real" job in corporate America. This job, while a great learning experience, provided the impetus for how I approach work now. Yes, I worked the 60–80 hours a week like everyone else, but I saw the toll this took on others before I noticed how it was affecting me.
>
> Due to the long hours and travel I didn't exercise or get outside enough, and missed a lot of social events. This was bad enough, but

then I watched a long-term, dedicated employee have a heart attack in the office. He lived, but needed time to recover. The company "discontinued" his job during the recovery process. This poor guy didn't have anything else in his life. I decided then and there I was out of there.

Ever since, my work, physical health, and personal life have been more in balance. There are occasions when work dominates, such as finalizing a dissertation, edits, etc. But I always know the end date. At this point if things are out of balance it's due to something in my personal life, not work.

Because of the powerful life events our colleagues have witnessed, they are now able to be much clearer about how to achieve harmony across the different parts of their lives. They embody a perspective that is about making purposeful choices based on a vision for their lives that is not predicated solely on "busy-ness" or professional accomplishments.

In other words, rather than provide some kind of decision matrix or pat advice on time management, the main goal of this chapter is on *conscious decision-making*. This is how we define "vision." If we give any advice here, it is this: first take the time to *consciously decide* what you want, to *set a vision* for yourself and your professional and personal life. Then, *practice* making choices in line with that vision. And finally, be willing to revisit and *revise* (which means to "see" again, to clarify) your vision frequently in order to refine it. So, if we are giving you steps to follow, it is these three:

(1) Set a vision for your life.
(2) Make choices that support your vision.
(3) Periodically refine your vision to make sure you are building the life you want.

That's it. We believe that any decisions, challenges, or conflicts you face can be resolved by taking the time to get quiet, listen to yourself, and answer these three questions.

Furthermore – and we address this at greater length later – we believe in setting goals for your *whole* life, not just your professional life. Professional and personal goals should be constructed alongside each other, in conversation with one another, because we are *whole* people. Compartmentalization, though effective by some measures, typically robs us of joy and fulfillment. It is better to see our lives in holistic terms, especially because we often have to make decisions that affect the many areas of our lives at once.

The reason we define this process as "vision"-setting is because, for us, it requires you to step back and take a holistic view of your life. It also is not something you do once – you return to it again and again, to make sure your have set your compass in a way that is effectively navigating you toward the vision you want for your life. You may have heard of the idea of having a "meditation practice" or a "spiritual practice," and vision-setting is like that – it is a kind of professional and personal habit, a periodic and repeated imagining that will allow you to establish a more fulfilling, effective, and joyful life.

SETTING A VISION FOR YOUR LIFE

We begin by acknowledging that it can seem as if there are many things that can get in the way of putting your vision for your life into practice. In the previous chapter we described the importance of having the courage to self-reflect and then act on what you know to be right, and we think that goal-setting can be one tool for getting clarity on the direction in which to move. But this is not enough; our goals are usually not reached by simply articulating them. In the face of all the outside forces in our life – activities that demand attention, other people who want us to move in other directions, an employment system that values more work for less compensation – it can be hard to stay focused and work toward the realization of our goals. In fact, the push and pull of outside forces can be so strong that it is easy to succumb and lose sight altogether of our goals and ideals.

But there is another reason why we might lose the direction in which we want to move, and that has to do with how we define "good work." We get the idea of "good work" from renowned social psychologist Howard Gardner, who defines it as a type of activity that is meaningful, purposeful, and ethical. It is this innate sense of good work that gives us the passion and dedication to turn a job into good work. Yet, according to Gardner (2008), many young workers find that pursuing good work is a luxury they cannot afford. How come?

There are a number of systemic explanations for why "good work" can be hard to find or practice. Employers may set limits on worker freedoms, compensation, and creativity, for example – often in the name of efficiency or control. For example, a teacher may have to teach such large groups that the quality of her teaching suffers, often to the detriment of students. In education, increasing attention is paid to numbers, outcomes, assessment, and achievement – anything that can be counted or measured as evidence of success or failure. And our language increasingly reflects this emphasis on quantitative measurement: in academia, for example, we persistently speak about "teaching load." This harmless-sounding phrase reflects the belief that teaching is a burden, a secondary activity that distracts us from the real work, which is research. Indeed, many universities and search committees (implicitly, if not explicitly) value research over teaching. It is no wonder that in such an environment a young professional, for whom educating young people is the good work gets discouraged from spending too much time on teaching. How to respond to these systemic issues is beyond the scope of this book, but we would just briefly suggest that, increasingly, faculty members could better collaborate and coordinate in order to resist these movements toward over-quantification and metrics, and articulate for themselves and others what good work can look like. Our educational system is crying out for such voices.

Carrying a teaching load by working on the conveyer belt of education.

We might also sink our own boats, though, when it comes to pursuing good work. We might decide that the deck is too stacked against us, or that we are powerless in the face of systemic changes. As a result, we could decide never to do the work of setting a vision for our lives to begin with. Unfortunately, if we have not set a vision for what we want or how we want to live, we could end up taking on work that we don't need to do or want to do, or that is in conflict with our core values. This reinforces our original belief that we are powerless to change our lives. We are easily tossed around by the desires of others, by changing metrics, or the call of busy-ness. We may feel unmoored, as if our days are not our own, and that we have lost control of where our life is headed. The truth is that all of us feel this way at some time or another. But putting in place a practice of vision-setting and gentle self-corrections can make these experiences less frequent and traumatic.

So what does it mean to set a vision for our lives? If you are someone who likes detailed steps for this kind of work, you may want to make lists of goals, both short-term and long-term, for the different areas of your life. In her book *The Joyful Professor*, Barbara Minsker (2010) encourages you to make goals for "major projects" in both professional and personal areas. Reading her short book might be a useful option for you.

But we believe that vision-setting really requires only that you do two things:

(1) Get quiet.
(2) Write down what you want for your life.

Step 1: Get quiet. This can be surprisingly hard to do given the persistence of the stories that we continuously tell ourselves. We call this the internal dialogue in Chapter 5 on Listening. Being quiet can be particularly challenging when you are in the habit of feeling busy or being busy or wanting to look as if you are busy. If we are honest, we can acknowledge that our ego gets rewarded for busy-ness. We want people to think that we are working hard, taking our jobs seriously, and that we are important (being occupied with tasks means that we matter). Taking time out of each day to sit and *do nothing* can bring up all sorts of resistance from the part of ourselves that values being busy or being seen as busy. We might label "empty" time as laziness, wasted time, or navel-gazing – there are all sorts of ways to dismiss this kind of practice. This quiet or empty time is often referred to as meditation, but in case you don't like this age-old (or new age) term, it is just about quieting the chatter of the mind.

Yet our personal experience – and the experience of many others – suggests that this quiet time can *fundamentally shift* the way we experience our days, time, our work lives, and our relationships. It can put us back in touch with a sense of who we are and what we really want, as opposed to what other people want (or we think they want!). It is a way of slowing down and reclaiming time, and of recalibrating the energy we take into the world. We remember to breathe. We remember our humanity, that we are not productivity robots. We may remember our sense of humor. Or our sadness. We can get some distance from our worries, or the many pressures that feel so real and demanding or suffocating. At the very least, we get a moment of distance away from our work, which will allow us to return to that work refreshed.

Being quiet is a way of slowing down and reclaiming time, and of recalibrating the energy we take into the world.

Exercise: getting quiet

If you are anything like us, you may feel tempted to skip this exercise entirely. Because, you think, nothing (no thing) will come from it. No journal entry, no next step, no deliverable. You are busy and don't have time to just sit. Plus: boring.

Well, we would argue that you are *exactly* the person who might benefit from this exercise. Here's what we want you to do:

(1) Using a timer (on your smartphone, watch, egg timer, stove, computer, whatever) set an alarm to go off in five minutes. We want you to sit quietly for the entire time, so checking to see what time it is will be disruptive and distracting. Let a gizmo do the counting for you.

(2) Sit in a chair, lay on the floor, stand with your eyes closed, whatever. Just get yourself into a comfortable position. And make sure you won't be interrupted – put a sign on your door and silence the ringer on your computer and phone.

(3) Now breathe. As you inhale, you can count your breaths. Or you can say to yourself "breathing in" as you inhale, and "breathing out" as you exhale, over and over again. Your mind will wander. Notice

> when it does, and come back to your counting or words (breathing in, breathing out).
>
> (4) The alarm will go off (we promise) and you'll be done.
>
> That's all you have to do. You just got quiet.

Maybe you have a strategy for getting quiet that works better than this. That's fine, too. Counting breaths is just an example. Whatever approach you pick, try sticking with it for at least four consecutive days, the time it takes to set a new habit, and see if you notice any changes (Beck, 2008). Do you feel calmer? Did you fall asleep each time you tried to breathe? Did you feel irritable about not working during the five minutes? This is all interesting information for you to have! If you feel calmer, perhaps getting quiet is a practice you will want to incorporate on a daily basis. If you fell asleep, that's probably giving you the information that you are overtired and need to get more rest if your life is to work better for you. If you got irritated about not being able to work during that time, then you could interrogate whether you are suffering with some workaholism and your life isn't functioning in harmony. Or maybe you are afraid of getting quiet because it forces you to confront some of your fears about how others see you, or about whether or not you are really "successful." All good information to have before moving on to Step 2.

Step 2: Write down what you want for your life. You can think of this as goal-setting writ large, if that resonates with you. Ultimately, our choices of what we take on, and what we don't, however, should be determined by the vision we have set ourselves. Setting goals can be more overwhelming and can result in longer to-do lists if we don't know what we really want. Without having clarity on your overall vision it is obviously impossible to use goals as a tool for making decisions about which activities to undertake.

A different approach – one that might precede determining individual goals – is actually *envisioning your ideal life*. To pose this as a question: if everything was just as you wanted it, what would your life look like? This is different from goal-setting, which implies a series of concrete steps that you will take to get where you want to be. Vision-setting is different: it is a way of imagining *how* you want to be. For example, a goal might be:

I will submit two articles for peer-review this academic year.

If that is your goal, you can make some choices each day to support that goal, such as blocking out a set writing time every day, saying no to this committee or that to protect that time, attending an article-writing workshop, and so on.

A vision, on the other hand, might look something like this:

I love having time each day to read or write, mentor students, work with colleagues who inspire me, have uncontaminated family time, and get some kind of exercise outside. I see each day as an opportunity to engage in the activities I love.

This is actually Jen's vision, and she uses it to set goals and also to prioritize the decisions she makes on a daily basis. This doesn't mean that Jen always gets to do *only* these things, everyday. Sometimes she is in long meetings, or is traveling, or has to care for a loved one, or gets sick. Sometimes a book manuscript is due and some of these other activities have to be put on hold. But overall, this vision – which articulates what Jen really desires from her life, and what brings her joy – allows her to structure her days so that she is engaging in these activities. This vision also leads to goals – because Jen makes time to read or write every day, she is always making progress on publications, which allows her to meet one of the externally established metrics for success: publishing. But it feels very different to come to the page every day because one wants to, rather than because one *has* to.

Exercise: write down what you want for your life

On one sheet of paper, take some time to lay out your long- and short-term goals, like Minsker (2010) suggests in *The Joyful Professor*. Doing this will give you a good sense of what you inherently value in your day-to-day life, and which external pressures you respond to. This is a kind of "brain dump" that will allow you to clear away all the "tasks" you feel pressing so that you can do the more big-picture thinking required for vision-setting.

Now, on another sheet of paper, write the words "My vision for my life is . . . " or "I see each day as an opportunity to . . . " and then complete the sentence. You can be as detailed as you like, but we tend to think simpler visions stick with us better throughout the day.

We also believe in the importance of *writing this vision down*. Commitment sounds great when it is just an idea or ambition that floats around in our thoughts, but can remain ethereal and abstract. In practice, our commitment firms up when we make it more tangible. We can do so by translating this commitment into words. This can be done, for example, by writing down what we have set out to do and hanging that text above our desk or on another visible place.

A final word: this is a living document, and it is totally specific to you. You can share it with others if you like, but really this is an articulation of your vision, for you, and you are welcome to protect it or keep it private if you wish. You might find your statement changes over time as you continue to reflect and come to know yourself better, or discover new strengths.

MAKE CHOICES THAT SUPPORT YOUR VISION

We like the practice of vision-setting because it provides a *guided flexibility* that we think can be especially useful to academics. If my goal is simply to publish two academic articles this year, that makes it difficult to decide whether or not to serve on a particular committee, or make time to go out to lunch with a colleague, skip a meeting to go to a child's dance performance, etc. However, if my vision involves simply making time each day to do some writing, to have meaningful interactions at work, to move my body, and to make time for my family, then that does indeed help me to make these decisions. I can ask myself: am I serving on this committee because it's an opportunity to work on a problem I care about, or with people I enjoy spending time with? If not, you can suggest to your supervisor a different committee assignment where the answer to this will be "yes." And maybe also going out to lunch with a colleague will offer an opportunity to connect and establish a meaningful collaboration, which might enrich your research or teaching. Or perhaps that colleague is a complainer who saps your energy, and you'd rather spend your time today at lunch working on your article.

Either way, having a vision for what you *want* as opposed to what you must achieve, can make these day-to-day decisions easier. Vision – defined as a continued practice of making meaningful choices in line with what we value – requires that we stay true to ourselves. This means that we don't let go of our goals and ideals, or our vision for our life. It also means that we carry out the pursuit of our goals in a way that is congruent with our values. It is this type of vision that allows us to know the difference between a calling and a career, as mentioned in the quote at the start of this chapter.

We need to know what our vision for our lives is because the modern world has become increasingly busy. In practice we encounter frequent demands on our attention. In many workplaces the pace at which work is to be completed is increasing, as is the drive to be productive. Universities increasingly measure output quantitatively in terms of number of publications, the amount of research funding that is generated, and the number of classes taught. These measures are relevant, but they are not a measure of the level of innovation of our research, or of the way in which we have helped students grow. These quantitative measures of productivity can easily prompt faculty to be overly focused on production instead of creative discovery and mindful teaching.

Furthermore, the continuous onslaught of new demands, new information, and new opportunities can easily divert us from doing the good work. Modern electronics allow us stay in touch with an unprecedented number of people, and this contact happens at a timescale that is steadily getting shorter. Complaints that there is "not enough time," or that "I am so busy that that I cannot do the things I ought to do" are common in the conversations of modern professionals. Many of us feel like hamsters in a wheel; we are running faster and faster and yet are staying at the same place. In this environment it is a challenge to be persistent and remain focused on the goals that we have set ourselves. As Dag Hammerskjöld (1964) writes in his book *Markings*: "Never look down to test the ground before taking your next step: only he who keeps his eye fixed on the horizon will find the right road."

Never look down to test the ground before taking your next step: only he who keeps his eye fixed on the horizon will find the right road.

How can we use vision-setting to make practical day-to-day decisions? What do we do when we are overwhelmed with duties and tasks, or conversely when we're excited about doing everything in our lives and can't prioritize? You've probably guessed by now that this is not a traditional book on "time management." In fact, there is no time to manage. We can only manage our activities; hence we speak of "activity management." We hope we have provided a simple approach to vision-setting, and choosing activities that will allow you to persist with your good work over time. Persisting in our good work, however, means that occasionally we must say "no" to that which doesn't serve our vision.

Our colleagues reiterated this point to us several times in our interviews with them. Knowing what you want (your vision) allows you to figure out what to say no to. One put it this way:

> What I've learned from working part time is that it is possible to spend a very limited amount of time working, while still achieving something worthwhile. There are tricks to this. One is limiting one's focus. You have to say "no" to many opportunities to do worthwhile and interesting work. By concentrating on a task or two, one can still achieve a "critical mass" effort on those problems.

Another colleague, when asked about the main challenges graduate students or early-career faculty face, wrote,

> I think it's that notion of FOMO (fear of missing out). When I was a new faculty member, I felt like I needed to respond to every proposal call; big and small. Then you inevitably miss a deadline or submit a proposal or paper that you weren't "into." I also see far too much perfectionism in my colleagues and students, particularly young female scientists. I've really tried to coach younger folks in the art of "fake it till you make it" and "done is better than perfect" but it's easier said than done.

Both of these colleagues are essentially making our point: knowing who you are and what you want is a crucial part of being able to prioritize the projects and activities that are most likely to lead to both joy and success.

When we suggest saying "no" more often, we don't want to advocate an asocial stance where we don't care about the needs of others. Instead, we learn how to say no from a balanced evaluation of our own needs and desires, as well as the needs of those around us, or of the organization in which we work. The mindset in which we are aware that we are part of a network with others, and that responsible choices involve our own wishes as well as those of others, is what Stephen Covey (1990) in his book *The 7 Habits of Highly Effective People* calls "interdependence." When we are dependent, we are driven by the needs of others. When we are independent, we are purely driven by our own desires. But it is in interdependence that our choices our driven by a deep understanding that we are part of a network. When we are dependent we cannot say "no" because we don't know how to do this or we lack the courage to do so. When we are independent we say no for selfish reasons. But it is in interdependence that we know how to say either yes or no, and know that we do so for the right reasons. Activity management thus is closely connected to our personal growth from dependence through independence to interdependence (Covey, 1990). The beauty of this is that it is in interdependence that life attains meaning, as stated in the following words from the book *Prisoners of our Thoughts* by Alex Pattakos, who built upon the work of concentration camp survivor Viktor Frankl: "Whenever we stop long enough to connect to ourselves, to our environment, to those with whom we work, to the task before us, to the extraordinary interdependence that is always part of our lives, we experience meaning. Meaning is who we are in this world. And it is the world that graces us with meaning." (Pattakos, 2008).

Why saying "no" might be difficult

For many of us, saying no proves incredibly difficult. For example, Jen works with a graduate student who is widely seen as highly skilled, gracious, and collegial. Yet this student is having tremendous difficulty making progress on designing and completing his dissertation, largely because he receives a great number of offers to do work that he is able to do well, and receives accolades for. Finishing the PhD feels both challenging and scary for this student (as it was for many of us), so he often says yes to other tasks or work that he knows he is able to do well instead. This leads to further feelings of frustration and inadequacy.

Jen asked this student to do the visioning work described earlier. This student would benefit from more clearly articulating a vision for his life, so that he can (1) either decide that he does not want to finish his PhD program, and would instead prefer to continue working effectively in the private sector, as he has been, or (2) decide that finishing his PhD is essential to realizing his vision for his life, and begin to say no to other work offers so that he can finish his dissertation. Indeed, having that clearly stated vision to guide his day-to-day goal-setting has clarified a seemingly complex set of choices for this student.

Like this student, many of us have developed feelings or thoughts that keep us from effectively saying no. We describe some of these here.

"I don't like confrontations or don't want to disappoint others." Saying "yes" obviously presents us with the path of least resistance, particularly for junior faculty members, who need to be seen as team players for political reasons, or just have personalities that make us want to please others. This could lead us to being overworked or unable to prioritize work (whether personal or professional). In our experience this leads to feelings of resentment. When we feel resentful, it becomes increasingly difficult to make decisions in line with our vision for our

lives – we feel trapped in a cycle of saying "yes," and don't know how to say "no" without becoming emotional or projecting our anger or frustration of being overwhelmed onto others. When we find ourselves in this pattern, we may need to return to our vision-setting and institute more opportunities for quiet time. We can also practice conversations where we can say "no" without being confrontational or resentful; sometimes writing possible responses down – such as "I'm honored to be asked, but my focus is really on [. . .]" – is useful.

"I like it when I am needed or when others think I am important." We can easily feel flattered when we are asked to participate in or take on a task. It makes us feel visible and valued. For example, we might feel honored to serve on a board or committee, or to be the colleague our Department Chair always comes to when she needs a sounding board. Unfortunately, the ego often cares about the short-term adrenaline rush it gets from being seen, and makes it easy to forget about our vision for our lives. As a result we may be tempted to accept a new activity that does not fit who we want to be or what we want to do, at the expense of spending sufficient time on matters that are more important. To make wise decisions we must learn to differentiate between the demands of the ego and the goals that truly matter to us. This is hard, and takes practice. But having a simple, clear vision can make it easier.

"If I am to survive I need to work as hard as I can." When we view our career as being driven by competition, or when we feel that professional survival is only for the fittest, we can easily lull ourselves into believing that we can contribute only when we give too much of ourselves – so much so that we become sick, tired, or resentful. This belief may be driven by a sense of guilt (perhaps you secretly believe you are lazy, or that downtime will be perceived as such by others), which tends to reinforce the belief that we need to work more. Or perhaps you are worried that your worth is defined mostly by your busy-ness at work, and if you are not busy, it is hard to know who you are, or whether you really are worthwhile. Such

beliefs can be so convincing that we may work at a pace that damages our health, our relations, and paradoxically the quality of the work that we deliver. If this is our mindset, then we need to learn that "more is less."

"I have so many things that I should do." Some of us feel that there are things that we simply "should" do. We may feel we have to attend every scientific conference, that we need to publish ten papers per year, that on the weekends we have to work, and that on top of this we also have to be loving parents or partners to our significant others. There are, of course, things that we "should" do; some duties are an inherent part of interdependence and accepting the responsibility that comes with it. "Caring for my children and/or aging parents" might be part of your life vision. But when too much of our life is driven by the "shoulds," we might not be able to devote ourselves to the things that matter most to us. In that case we might be served by a careful evaluation determining which of the "shoulds" can be abandoned.

If you find it difficult to say "no," it can be worthwhile to spend some time reflecting on the underlying reasons for why this is difficult. A better understanding of your own barriers to saying "no" might make it easier to make choices that create space for the issues that are most important to you. You could do this reflection on your own, but it may also be helpful to do this together with a trusted friend. It is helpful to support each other in identifying what to say "no" to. Margaret Palmer, a prominent environmental scientist, has started a NO-club (Stokstad, 2014). The club consists of a group of trusted colleagues that meet weekly. Members of the NO-club have agreed to discuss any decision on a new activity in the NO-club before committing. Your colleagues in the club can ask you questions that provide valuable insights; they may ask you to articulate how your choice is aligned with your goals and vision (or not), and they can provide mental support for saying NO when that is the right decision. The NO-club thus combines a reality check and moral support.

Exercise: say no

Perhaps you are a person who finds it difficult to say no. Indeed, it's difficult to figure when to say no if you don't have your vision in place. Once you have done the vision work described earlier, we suggest that you practice saying no to activities that don't fit your vision for your life or satisfy the requirements of interdependence.

Begin by developing a non-confrontational phrase and tone you can use for when you say no. You can try this one, for example: "I understand that this task is important, and am grateful you asked. But I am putting my energy on completing another project right now." Notice what happens inside yourself when you do this, and notice what reactions you get from others. If this is difficult for you, it may help to figure out what the consequences would be of not saying no, and who would pay the price that you had taken on too much. Use this exercise to build up a certain level of comfort in saying no. As your comfort level increases, you can gradually be more daring in what and who you say no to.

REFINE YOUR VISION

The concept of the reality check brings us to the third step of realizing your vision: periodically refining your vision to make sure you are living the life you want. In the face of the distractions of every day it can be a challenge to maintain focus on what you really want. Without such focus it is difficult to muster the persistence needed to move forward. Posting goals in a visible way provides the reminders needed for staying focused. But perhaps the most essential step in maintaining focus is to go back to your vision whenever a new activity presents itself. In that case it is essential to ask yourself the question *does this new activity support my vision for my life?* If it does, it likely is worthwhile to accept the new activity. If it does not support your goals, you might still take it on, but there must be good reasons for doing so, such as satisfying requirements for interdependence. Having clarity on these reasons is essential for making good decisions.

But our visions for our lives can readily shift over time. It is helpful to think of our life as a collection of different "seasons" – phases or stages at which our needs, commitments, and wants are determined by our networks, age, abilities, and challenges. For example, the "season" of beginning a new career may bring with it a vision that looks like this:

> My vision for my life is that I am surrounded by mentors and colleagues who help me to become a better researcher and teacher, I successfully communicate my goals to my colleagues and superiors, and I maintain a healthy lifestyle with enough rest, exercise, and relaxation.

For a faculty member who is in a different "season" of his life – such as mid- to late-career, a vision might look something like this:

> I see each day as an opportunity to enable my students and junior colleagues to successfully reach their goals, to work on projects that are innovative, meaningful or joyful, and to participate in activities where I can spend time with my loved ones.

Statements like these will be different for every person, and each individual person could ostensibly have multiple life visions over the course of his or her life, dictated by our changings needs and desires. Disabilities, career changes, the arrival of children, the ebb and flow of responsibilities, promotion, epiphanies, all provide us with opportunities to revisit our vision for our lives so that it resonates with the "season" we are living in. Allowing yourself this flexibility and encouraging yourself to update your vision will also give you compassion for others as they go through the various "seasons" of their lives as well.

You may also find that some part of your vision is not actually working for you in practice, and needs to be adjusted. One of the colleagues we interviewed noticed that one of the strengths he had – collaborating with and supporting others – satisfied his need for cooperative partnership, but also meant he did not make enough time for his own work. He wrote,

> I am a collaborator, which is also evidenced by the fact that I have only one paper with me as single author. This means of course that my personal signature in science is not very strong and I am at peace with that. In my experience, being a valued scientist does not strongly depend on it. I am compassionate and enthusiastic, but not good in saying no to anyone who comes to me with a question or an idea. This means that I am deeply embedded in the professional lives of others, which is good for my feeling of connectedness, but it also means that at times I struggle with my time.

For this colleague, collaboration is a key part of his self-knowledge and a part of the vision he has for his life. Yet he also told us that he has to be aware of when he is paying too much attention to others and not enough to his own work, and make adjustments as he goes. Our hope is that the vision process provides opportunities to notice when these imbalances occur, and to correct for them.

Effectively implementing one's vision, therefore, requires an open mind and reflection to be effective. Reflection allows us to *reflect back* on ourselves, and invites us to contemplate what life we are building. Are we living a life in line with our values? Reflection allows us to consider such a question. True reflection requires an open mind, for it is only with an open mind that we can see an accurate image of ourselves, a reflection that is not tainted by prejudice or hidden assumptions. Persistence makes us determined to stay the course, while reflection makes us honestly evaluate whether we are actually staying the course and what course corrections are needed.

Our hope is that reflection ends up being a largely positive experience – it is a chance for you to give yourself some honest feedback and to periodically revise your vision for your life, which then allows you to make better decisions. All of this should lead to more joy! But we are also aware that many of us, and perhaps academics especially, can be particularly harsh judges of themselves. If we've encountered a failure, challenges, or a series of setbacks, it can start to feel as if something is wrong with us, or as if we are being punished in some way.

If this is the case for you – and we all have seasons of our lives in which we've struggled – pay particular attention to the language you are using when you reflect. Get quiet, and notice which thoughts arise as you contemplate your life and your vision. Do you say to yourself, "I should do a better job staying the course! I haven't met any of my goals. I'm failing!" Or do you use blaming language to fault others? Do you say, "Why am I always being pushed in a different direction than I want?" Notice when you are using harsh or judgmental language toward yourself or others. Then reflect on whether or not you have articulated the right vision for yourself, and whether your actions are in line with that vision. Having alignment between vision and action, in our view, is often a sure road to both joy and success.

Indeed, we hope that you aim to reflect and revise your vision for your life not from a place of negativity or judgment but from a place of compassion and love for yourself and for what you want out of your life. This may sound a little out there, but we believe that you cannot persist in building the life you want unless you do so from a place of love. Persisting from shame or guilt or fear may lead you to meeting some goals for "success," but that does not mean you have built the life you wanted along the way. You may have to pay attention to how you *feel* when you are doing your reflecting and revising, as opposed to simply plugging in what you think the "right" answers are. Perhaps you *think* you must get three new grants this academic year. Your brain tells you that is what is necessary for "success." But maybe you have a loved one you are caring for, and you have decided that part of your life vision is to practice yoga regularly and to do some reading for fun. These things are examples of activities that may *feel* right to you.

So perhaps you will have to modulate the number of grants you will seek. In this way, you can avoid judging yourself when you don't meet that unattainable funding goal as well, and can work more productively on the grants you *do* get because you are feeling healthy, better rested, and able to take care of your parent in a way that brings you a sense of well-being and fulfillment.

Another benefit to this gentler approach is that the less we judge ourselves, the more compassionate we can be toward others. When we set a vision for our lives that allows for us to harmonize the many parts of our life, we can also support that for others. If we don't allow ourselves any free time or exercise or fresh air or time with loved ones, we most likely begrudge that for others as well. We can probably all identify situations in which we've judged others based on partial evidence or our own harsh expectations for ourselves. For example, we might resent a colleague who never speaks up at meetings because we presume he has no opinion and does not want to take responsibility, while in reality this colleague is just introverted and simply does not get the space an introvert needs to speak up (Cain, 2013). If you are someone who tends to make a lot of assumptions about the motivation or performance of others, this may be something you want to reflect on. In this case "practicing compassion for myself and others" could become part of your vision for your life.

We also note that in setting this vision for your life, and in practicing quiet reflection and revision, some of you (or many of you) may encounter resistance to these practices. We certainly have. There have been times when the habits of busy-ness, cynicism, or distraction have overwhelmed us as well. We think that resistance is a natural part of this process, and we invite you to notice it and question it when it arises.

Exercise: when do I resist?

Rather than ignore your resistance to quiet, vision-setting, or reflection, respond to these questions in your journal:

(1) Why am I holding on to my old beliefs or habits about work, or my life, if they aren't working for me anymore?

(2) Frankly, I feel like this is a bunch of self-help mumbo-jumbo. People who do this kind of work just don't know how to succeed in real life. But, I'm going to pretend for a moment that there might be something to learn from this. If I set aside my disbelief, what might I have to gain from trying the reflection and vision-setting exercises? Am I afraid of losing anything by trying these practices out? What would my harshest critic say about my trying these activities? What would my most supportive friend say?

(3) I just can't make the whole "getting quiet" thing work for me. I'm always being interrupted and it just makes me antsy. What is it that bothers me about this practice so much? What does that resistance mean? Is there a way I can tweak this practice to make it work for me, such as trying a walking, instead of sitting, meditation?

Our point in asking you to bring your resistance into the light, so to speak, is to encourage you to find practices that work for you, and to help you to see yourself better. When we are unwilling to consider alternative options or solutions we may simply be stubborn, or we may have located a symbolic place in our life where we most need to change (we tend to believe the latter is true). The point is not to force yourself to do something you detest, but to use these practices as an opportunity to grow, to flex your reflection muscles until they are strong enough to serve you on a consistent basis.

On the other hand, perhaps you are consistently meeting your goals and feel as if your vision for your life is easily met, but you are feeling a bit "itchy" or bored with how things are going and you can't articulate why. In this case, we would ask you to consider whether you have set your own bar too low. Maybe you are someone who thrives on a bit of adventure, and that is not reflected in your life vision; in that case maybe you could add, "Once a semester, I will engage in a professional or personal activity that frightens me a little but which makes me feel alive." Then go ahead and schedule that adventure!

Once a semester, I will engage in a professional or personal activity that frightens me a little but which makes me feel alive.

To put this another way, sometimes we limit ourselves *too* much to what is "realistic," and we refuse to *dare*, to push ourselves out of our comfort zone. When skiing, Roel's children would sometimes point out that he had not fallen once during a day. Roel would be proud of his ability to avoid mishaps, but his children challenged him for never falling because it meant he wasn't pushing himself, he was not taking any risks. Not pushing himself meant he was not getting better at skiing. Roel's kids were right; it can be useful to formulate goals that are so ambitious that we occasionally fall and don't reach all of our goals.

The acts of bold visioning and risk-taking are closely intertwined. The more daring our vision is, the more willing we may be to take a risk. On the other hand, when we are more willing to accept risk, we are more open to new ideas that may change the way we do things. Imaginative visioning and risk-taking both involve letting go of control. The paradox is that the reduction in our creativity and imaginative power caused by a desire to control actually reduces the control over our life and the world around us, because with a reduced imagination we are less capable to give direction to our growth and development of new initiatives. Developing a clear and bold vision ultimately gives greater control because it is a driver of new directions to seek out. Conversely, control without vision seeks to maintain the status quo and is averse to change.

Our point is that our visions for our lives, and their accompanying goals, should always be open to adjustment. And we recommend scheduling these adjustments regularly, so that you are consciously revising your vision and your goals, rather than just allowing things to slip or slide as a result of the conditions in our life. Each of us is responsible for driving this process, and we recommend making it a conscious decision whether to change a goal so that we are aware of what we have changed and have clarity on why we have done so. Goals and vision should not be seen commandments that are carved in stone.

Self-reflection also requires a level of honesty with ourselves that some of us may struggle with. For example, say a junior faculty member is struggling with the tremendous pressures facing him and finds that while he can keep on top of emails and class prep, there is never any time left at the end of the day to write up his research. It is the end of his second year on the tenure-track and he is starting to panic about his lack of "productivity." What should he do?

Although this is not a book on time management per se, we might have two reactions to this young scientist's dilemma: one is that there is a structural problem – such as too many students in the classroom or too much service – that needs to be resolved through negotiation with the scientist's supervisor.

But we would also ask this faculty member to take a good, hard look at his own habits. We might ask this person to spend a week tracking *exactly* how he spends his time, and to identify behaviors or habits that aren't serving him in meeting the part of his vision that includes meaningful writing periods. Such a record might reveal that he is spending too much time on Facebook in between checking emails, or is consistently an hour late to the office everyday because he is staying up too late at night watching movies and can't get up early enough. Perhaps he is giving too much time to class prep, or needs to physically schedule uninterrupted blocks for writing into his calendar as visual reminders to meet this goal as part of his vision.

In her essay on working 40 hours a week as an academic, Roberts-Miller makes the point that, in practice, we tend to spend our time at work often not very effectively because we seek distractions, procrastinate, chat with others, surf the Internet, or don't focus on our work in other ways (Roberts-Miller, 2014). We readily accept such distractions, and in fact, we often encourage this behavior in others by being a source of distraction ourselves. It is a great paradox that while, on the one hand, we sometimes engage in ineffective work habits, we insist, on the other hand, that to be successful in academia one needs to make extremely long workweeks. We may have never broken grad school habits in which time and responsibilities were fluid, and personal and work time flowed together, either in the form of overwork or procrastination. There may be merit to some of these unproductive activities; when done in moderation it can be useful to chat to colleagues about personal matters or to binge-watch reality TV. But Roberts-Miller makes the point that time spent on non-essential activities must be compensated for by excessive working hours if the essential activities are to be carried out. For some of us, the long work hours we make may thus be caused by ineffective work habits. It could be that actually working 35–40 hours a week, during "regular" work hours, can be particularly effective. For others, the flexibility afforded us by the academic life – so we can be home when our kids get home from school, for example – may mean that a typical 40-hour workweek looks different. But working 60 hours to compensate for distractions, we would argue, means that most likely something is off kilter.

Our point is that there often is a discrepancy between the way we actually spend our time, and the way we think we spend our time. You can get insight whether that is the case for you by monitoring for a week how you actually spend your time. If you prefer, you can use one of the many electronic tools that is available to monitor your activities. A comparison between the way you actually spend your time and the way that you think you spend out time can give important insights into your work habits. If, after doing this, you find that you really do not have enough time to live your life vision as you've articulated it, then you need to revisit the goals you've made, say no to some things and yes to others, and try again. This is why the revision process is so important – it allows us to adjust things that aren't working and puts us back in charge of our lives.

In conclusion, this process – the practice of vision-setting – is one of trial and error, reflection and revision. Our vision needs to be adaptive so that we can change it when we discover what works and what does not work. This adaptation is most easily done by setting up a cycle of planning, evaluation, and revision. You don't need to be alone in this process; it usually helps to get the input from others, especially those who have experience with the type of goal we would like to reach. Don't hesitate to seek the help and advice from others!

Exercise: monitor how you spend your time

We often spend our time in very different ways than we think. For this reason we suggest in this exercise that you monitor how you spend your time for a week. Since professional and personal lives both take time, it is important to cover both. So write down items such as Monday: sleep (7 hours), email (1.5 hours), reading scientific literature (1 hour), etc. Add the amount of time spent on different activities. Next, write down what you would have liked to do, but what you could not find time for.

This is the easy part, now comes the hard question: what it is that you want to change? Write down what changes you want to make to the way you spend your time, and integrate these changes into your vision.

4   Curiosity

I saw the twinkle of mischief in their eyes.

I saw the smiles of triumph and elation as a new task was mastered.

I saw their "wicked" grin, full of playfulness.

– Boudewijn Roodenburg-Vermaat

These lines are the opening lines of a letter written by Roel's cousin Boudewijn. Boudewijn has always had a great passion for the outdoors and the rural life. He also has a deep desire to work with disabled people. Since he is a person that follows his passions, he worked for many years on a farm where a number of mentally disabled children lived with him and his family.

The lines above express the deep love Boudewijn has for his "pupils," as he calls them. But these lines also powerfully express a joy of life in these children, including their joy of learning and an irresistible urge to play. The "twinkle of mischief" in their eyes refers to a spark, a beam of light, something that both illuminates and prompts unexpected action or "mischief." This type of mischief ignores limitations in the form of rules set by others or by "common" knowledge. Boudewijn also mentions the children's triumph and elation as a new task is mastered. They have a drive to learn and experience deep joy from mastery. This joy is not only fueled by the satisfaction of learning a new thing, but perhaps more importantly it comes from a sense of accomplishment from going beyond previous limitations. The children's "wicked" grins are playful and suggest the willingness to do something with abandon, to be completely absorbed by it.

This is what children do: they explore, they play, they locate boundaries, and then go beyond them. Children are the natural explorers of the world. We tell this story about Boudewijn's pupils because it beautifully encapsulates why we believe curiosity is such an important character trait for scientists. Curiosity, as we think of it, is fundamentally a form of *delight*. When practiced as a form of delight, curiosity also becomes an effective antidote for the most common ills that beset the academic scientist: getting stuck, feeling anxious, and feeling isolated. In this chapter, we describe how to recapture a sense of play and delight through practicing curiosity, and we also suggest that if you are suffering from any of the common difficulties listed above, getting curious may be just the thing you need.

CURIOSITY AS PLAY

Many of us lose the sense of playfulness, the drive to explore, described above, as we get older. We feel overwhelmed by our many obligations, and that play is a waste of time. Some of us have *never* been playful, in the stereotypical sense of the word – perhaps you have met a "serious" child who is something of an "old soul," who seems mature beyond their years. It is also possible that many of us were educated in a system that squeezes many children into a mold in which they do not belong. This certainly was the case with Roel's cousin Boudewijn, who thrived and was filled with purpose, but only after he could leave school himself. In his TED Lecture *How Schools Kill Creativity*, Ken Robinson describes the squelching of creativity by an educational system geared to serve the needs of the industrial world (Robinson, 2006). Many of us give up our inherent playfulness after we are told too many times to "not be silly" and to "grow up." A healthy dose of realism can, of course, be a useful correction at times, but when overdone it can extinguish our creative sparks, our desire to learn more, and our experience of life as an opportunity to play.

We encourage the reader to develop an "itch to know." When something itches, we have an instinctive urge to scratch, to do something about the itch. In the same way, an itch to know brings an instinctive urge to pose research questions and to seek answer to these questions. How do we nurture this itch to know?

- *Lean into a sense of wonder.* We described above how powerful a sense of wonder can be, and how natural this sense of wonder is for children. But as adults, we are tempted to subdue this sense of wonder by being overly rational or skeptical. Leaning in to the sense of wonder naturally fuels the desire to know.
- *Cultivate the mindset of an explorer.* Men like Vasco da Gama, who was the first European to sail to India, went beyond known boundaries and ventured into the unknown. This is what explorers do, and this is what good researchers do. The mindset of the explorer is the mindset of somebody who wants to know.
- *Jump-start something new.* When we work on a topic for a long time, we know the topic so well that the curiosity and passion may wane. Switching to a new line of research or teaching can create a new sense of wonder that rekindles curiosity and passion.

These pieces of advice may sound easy, but in the face of the everyday pressures of the research environment, it can take a conscious effort to lean into the wonder, to have the mindset of the explorer, and to seek a new line of work.

Exercise: how is the child in you doing?

This description of the mindset of children begs the question: is that childlike spirit to learn, play, and explore alive and well in you? We suggest you spend some time reflecting on the following questions:

- What are the activities that feel like "play" to you?
- What gives you a sense of wonder?
- If the notion of "play" doesn't appeal to you, think about what activities put you in "flow." What activities make time pass without notice?

Write down whatever comes to mind. There is no need to compose nicely flowing prose; jotting down keywords is enough. The main point is to record your thoughts and feelings and to use the action of writing to make your thoughts and feelings more specific.

In this chapter, we would like to offer a rich and broad definition of the notion of play, as it is operationalized through the practice of *curiosity*. What we mean when we talk about play is the willingness to explore, to be surprised, to get lost in doing something even if the outcomes or "deliverables" are not clear. We also believe that play involves a willingness to be wrong, to make mistakes, and even to appear foolish in front of others. The meaning of play can therefore be interpreted any number of ways. For some of us, play offers a respite from work. It allows us to engage parts of our brain we aren't typically using in the lab or the classroom. For example, play might look like mountain biking in the foothills, splashing in a pool with our children, playing Sudoku, or putting headphones on, lying on our backs, closing our eyes, and listening to an entire album of music – what a treat! In this way, play can feel like a kind of active rest, an engagement in activities that reanimate and rejuvenate us before our natural cycles take us back to work.

But play can also happen *in* the workplace. Again we refer to the concept of "flow," which means that we give ourselves over to whatever we are immersed in, so much so that time seems to cease passing and we exert much more energy and focus than we might normally (Csikszentmihalyi, 2008). When thought of this way, "play" can actually come to look a lot like "work," if work puts you in flow and leads you to generative discoveries, innovation, and a feeling of dynamic possibility. If you are someone who can spend hours in the lab without noticing that time has passed, or can spend hours on your computer working on a manuscript, or hours mentoring graduate students – and you are doing these things because you simply love doing them – these are healthy experiences of being in flow and experiencing a playful concept of work.

We think it's likely that many of you are intrinsically curious people, and that wanting to know more about the world and how it works is what brought you to science or engineering to begin with. Curiosity in research is usually driven by questions that come up persistently, questions that simply won't go away. In these cases, it

It almost feels not as if you are coming up with the questions, but that the questions come to you.

almost feels not as if you are coming up with the questions, but that the questions come to you. The central role of asking questions in research, and of using questions systematically to drive research, is addressed in detail by Snieder and Larner (2009).

But we also think that we sometimes lose sight of curiosity as we face the day-to-day pressures of being successful academic scientists. For example, most of us know that in order to do research it is often necessary to secure funding. The funding may be available on topics or for types of research that don't really excite us. Conversely, the research about which we are most curious might not immediately lead to a steady stream of publications, perhaps because that research is risky, or slow to produce publishable results. For faculty who are under pressure to publish as much as they can, this might be a deterrent for pursuing such an adventurous line of work. Or we might feel exhausted by the daily grind of answering emails, attending meetings, and administering funded research. It can be difficult to stay in touch with our feelings of curiosity if our day-to-day lives are set up so that we are just meeting one obligation after the other.

Furthermore, our egos can get in the way of us being curious: we are trained as scientists to "know" a lot about the world and to continue mastering new knowledge. We are also trained to demonstrate to others that we are knowledgeable. Always knowing the answer, always being right, being the first to speak rather than listen, perhaps condescending to others – all are habits that can contribute to our inability to connect meaningfully. Unfortunately, we think that many of these skills – connection, humility, and listening – are sometimes missing in academic scientists. Practicing curiosity, on the other hand, allows us to become much more effective in the workplace and at home.

We want to make the point, however, that being curious and playful is not merely practical: we do not rest or play only so that we can be better at work. We do these things because they are a way to express our care for and interconnections with the world around us. They satisfy a deep need in us to belong. In fact, the word "curiosity" comes from the Latin "curiosis," which means careful, from the root cura, to "care." Being curious, then means being "full of care" – we care enough about the world around us to ask questions about it, to engage in it. We are curious about the things for which we care deeply. We are curious about others, and we express that curiosity from a place of care and a desire to connect.

If you are struggling in areas of your life where you feel lonely, ineffective, or unsuccessful, it can be tempting to blame yourself or others, or to overthink and over-analyze. Addressing such problems with an attitude of curiosity, on the other hand, can lighten the load, allow you to develop a strong sense of self-awareness, and move you toward problem-solving strategies – including building deeper connections with people who can help you. Below, we discuss three common areas of struggle we have seen in our own lives and among our colleagues – feeling stuck, feeling anxious, and feeling isolated – as well as how curiosity might serve to alleviate them.

CURIOSITY AS ANTIDOTE TO FEELING STUCK

In our experience, feeling stuck – whether in our research or in our personal lives – is often a product of believing in certain fixed rules or constraints that we perceive to be immovable. Here is an example that illustrates what we mean:

> John is working on a first draft of an article that his advisor wants done by Friday (it's Monday!). John feels ill-equipped to write an academic article – this is his first year of graduate school – and his advisor seems too busy to teach him how to do it. John also feels like he should know how to do this – he's a PhD student after all, and everyone else seems to know what they are doing.
>
> John sets himself a rigorous writing schedule of working on the article for eight hours a day until it's done. He manages to write an introduction, which he thinks is terrible, but then he gets stuck on writing the literature review. There are so many articles to read, and he is not sure which are important. He spends the bulk of the first day printing out dozens of articles to read and ends up writing hardly anything. When he wakes on Tuesday, it is with a growing sense of dread. He looks warily at the stack of articles he's supposed to read, and then ends up spending several hours staring at the computer screen and wondering if he has made a terrible mistake by coming back to school.
>
> By Friday morning, John has only a draft introduction and a list of bullet points. He wakes with a terrible feeling of anxiety in his stomach regarding that afternoon's meeting with his advisor. He knows his advisor is sure to be angry, and to wonder what John is doing in the program.

John's situation – though hypothetical – has elements that each of us has experienced, or seen colleagues and students experience. There are so many assumptions embedded in the way this situation unfolds for John – so many unspoken rules that he feels bound by – that it is no wonder John ends up feeling stuck. This feeling of stuck-ness ultimately leads to intense feelings of failure and self-doubt for John.

But this is exactly the kind of experience that can be most helped by a detached sense of curiosity about what is going on, and about the implicit "rules" of conduct we feel ourselves to be bound by. Here are some of the hidden beliefs John might be operating under:

(1) I should know how to write an academic article.
(2) If I don't write this article correctly, people will know I don't belong here.
(3) I can't ask for help on this project, because if I do, people will find out I don't know what I'm doing.
(4) Articles can be written by first-year graduate students on their own, in a week.
(5) My advisor is going to be angry with me for failing.

You can probably think of others. Each of these beliefs leaves John feeling stuck in an unwinnable situation in which he is bound to fail.

Author Byron Katie proposes a remedy for such situations, which she calls "The Work" (Katie, 2015). In "The Work," we perform four steps interrogating whatever belief we have identified as keeping us from accomplishing the life we want to live. The Work is a great approach to applying curiosity to problem solving in our personal and professional lives. The Work involves four questions that you ask about your belief:

(1) Is it true?
(2) Can you absolutely know that it's true?
(3) How do you react – what happens – when you believe that it is true?
(4) Who would you be without that thought? (Katie, 2015)

Katie argues that this process of inquiry helps us to identify limiting beliefs that keep us trapped and to clarify that we actually have options for freeing ourselves from those situations.

Let's apply Katie's inquiry process to the third of John's limiting beliefs, listed above: "I can't ask for help on this project, because if I do, people will find out I don't know what I'm doing." First, we ask if that belief is true. John, of course, is going to say, "Heck yes, it's true! Look around me! Everyone here is busy working in their cubicles on their projects. Samantha published an article with her advisor last week and everyone has made a big deal about it. That's just where the bar is set."

Fair enough. It is possible that John works in a competitive environment where graduate students are expected to perform at a high level. Maybe people are not supportive or helpful in that environment. But we would encourage John to continue with inquiry nonetheless. The next step, then, is to ask whether we can *absolutely know* our belief is true. In response, John might pause. He might answer:

> Well, my advisor did encourage me to go to the writing center if I have questions, and he said I should talk to Samantha about her experiences. He told me a very rough draft would be acceptable, and that we'd work through it together once I had something down on paper. I see the other grad students talking to each other about their work a lot. I thought maybe they were just complaining to each other but maybe they could have some insight. I could also call my friend Cory, who's in a different program, and see what he thinks.

John's response to the second question, when compared with the first, represents a pretty big leap in terms of his hidden beliefs. In his first response, he is trapped in an unwinnable situation. In the second, we begin to see some lifelines or possibilities appear that might shift the situation for him.

The third step – how do I react when I believe my hidden belief is true? – is also key. For John, when he believes that he can't ask for help because it will reveal his ignorance, he feels small, frightened, ill-equipped, and alone. Clearly, the belief is not leading him to the right frame of mind for taking on a new, difficult task, such as drafting an academic article. The belief feels right and true because we often believe the worst things about ourselves – it's an odd form of self-protection. But that doesn't mean the belief actually *is* true, and if it's not getting us the results we want, it might be better to question the truth of it.

Fourth, Katie would ask John who he would be without the thought that he can't ask for help. John would feel less alone, like he has options, like there might be hope that he could get something ready to show his advisor by Friday, or he could ask for an extension. He would feel lighter, more optimistic, and more able, and he would maybe have begun building a sense of community with other graduate students who possibly are feeling the same things as he is.

A final step in Katie's The Work that is useful is called the "turn-around." In that step, we take our initial belief and we frame it differently, or pose it in the opposite fashion, to see if *that* belief feels better and might actually be true. For John, he might restate the belief to say, "I can ask for help, because I don't know what I'm doing (and that's just fine)." Or he could turn it around by saying, "I can ask for help, so that I can figure out how to do this better." Or, "People might respect me for taking responsibility by asking for help." Any of those beliefs leads John to feel better than the original hidden belief, and most likely will lead to John seeking and getting the help he needs.

Exercise: do The Work on one of your hidden beliefs

Write about a situation in which you feel frustrated or stuck. Journal for 5 to 10 minutes about the situation. When you've finished writing, go back and draft a short list of hidden beliefs you have about the situation – beliefs that absolutely seem true and right and fixed. Then do Katie's Work on one of those beliefs:

(1)   Is it true?
(2)   Can you absolutely know that it's true?
(3)   How do you react – what happens – when you believe that it is true?
(4)   Who would you be without that thought? (Katie, 2015)

Don't forget to do the "turn-around" at the end.

Now journal for a few minutes about what this process was like for you. Do you feel differently when you look at the turn-around beliefs you came up with? Does it suggest any new ways to act or react in your "stuck" situation?

CURIOSITY AS ANTIDOTE TO ANXIETY

Academics don't have the market cornered on anxiety; according to the Anxiety and Depression Association of America, anxiety disorders affect nearly 20 percent of the American population (Anxiety and Depression Association of America, 2015). But, despite being high achievers who are often responsible for public speaking, project management, and working collaboratively in groups, we believe that academics and scientists are probably just as prone to experiencing anxiety – if not anxiety disorders – as the general population. In this section, we'll speak about anxiety in general terms, as moments of anxiety, such as those caused by interpersonal conflict, and as generalized anxiety, such as that caused by prolonged feelings of inadequacy or unrelenting stress. We argue that some practices rooted in curiosity – much like The Work, described above – are possible solutions for both kinds of anxiety.

The kind of curiosity we're advocating here is a curiosity in yourself, a curiosity of the self-reflexive kind, although we believe this kind of reflexivity should be productive, and not merely navel-gazing. We also don't want to encourage an egocentric preoccupation with your self. But we do suggest that it may be helpful and illuminating to reflect on what choices you make, on how you act or react, and on what drives you. For such reflection to be useful, it is helpful to be curious about yourself, because this curiosity is likely to bring interesting insights to the surface.

The first kind of anxiety – what might be called situational anxiety – arises as a result of particular circumstances. Perhaps you have a colleague who one-ups you in conversations, or you have a student who disrespects you, or you are unable to meet a deadline. The inability to figure out how to deal effectively with these kinds of difficult situations can lead to anxiety.

The second kind of anxiety is perhaps more low-level or background than the first kind of anxiety, though it is also more pernicious in the long term. Rather than the peaks and troughs of situational anxiety, ongoing, low-level anxiety never really subsides; it acts like relentless static in the back of the academic's life. Occasionally, it peaks as deadlines or decision-points near, or when one receives negative feedback about one's performance. But for many, it never really goes away.

This kind of ongoing anxiety is both typical and harmful. It is typical because many academics feed on stress to keep going through long hours and difficult, long-term projects. Being an academic scientist also means that there is very often "bleed" between work and one's internal or home life – there is little "uncontaminated" time where one is not on call for work, or thinking about work. This kind of anxiety can lead to a refusal to take breaks for fear of falling behind, or it can lead to myopia about work, where that seems to be all that matters.

Ongoing anxiety is also dangerous because it can lead to burn-out and even impaired judgment. Unrealistic expectations about what can be achieved lead one to never feel satisfied about accomplishments. There is never an opportunity to experience what writer Anna Kunnecke calls the "gift of completion" – allowing yourself to savor and celebrate what you have accomplished or completed before going on to the next task or goal (Kunnecke, 2015). Instead, we focus immediately on what is *not done yet*, or we compare ourselves to what others have done, and we keep that dull roar of anxiety humming persistently in the background. We use it to fuel the next task.

The problem, of course, is that anxiety drains us instead of empowering us. As a result it can leave us feeling tired out, resentful, or scared. Not joyful at all.

Happily, both types of anxiety are amenable to practices of curiosity – in fact, we find that anxiety may disappear or resolve itself with gentle and genuine questioning. In her book *The Joy Diet*, author Martha Beck (2003) proposes the following questions, which encourage us to be curious about our experiences as a key element of experiencing more joy:

(1) Why am I avoiding stillness?
(2) What am I feeling?
(3) What hurts?
(4) What is the painful story I'm telling?
(5) Can I be sure my painful story is true?
(6) Is my painful story working?
(7) Can I think of another story that might work better?

If you've noticed some overlaps with Byron Katie's "The Work," you would be correct. But Beck's questions are especially effective for addressing anxiety because they encourage us to begin from a place of stillness and to pay attention to where the anxiety is lodged in our bodies. We find both to be particularly important practices for curiosity to do its magical work on anxiety. Anxiety only works if we remain on autopilot, refusing to get quiet and interrogate the internal stories we tell ourselves about what must be done and how. In this case we are easily overwhelmed by the babel of anxious voices within. Quiet, or stillness, works the other way – it encourages us to be present and to pay attention. And paying specific attention to where anxiety is located in our bodies forces us to acknowledge (1) that we have bodies, (2) that those bodies are giving us valuable feedback, often in the form of pain or tightness, and (3) that we can soothe ourselves through paying attention to calming the parts of ourselves that are worked up.

We can look at another hypothetical example to illustrate. Joanie, who is in her fourth year on the tenure track in Physics at a research-intensive university, experiences both situational and ongoing anxiety in her daily worklife and has become so disenchanted with the stressors of being an academic that she is considering leaving the tenure track. Her department has very high tenure and promotion expectations, provides little mentorship, and is also something of a "good old boys' club": she is frequently ignored or talked over at department meetings, and there have been some sexist comments made about women's abilities to succeed in her field. In addition, her mother is disabled and requires Joanie's significant attention to maintain the level of care she needs. Joanie is publishing articles, but worries that the journals she publishes in are not prestigious enough. She is also worried because she has not been applying for grants to fund her research, and worries that she doesn't have the skills she needs to be successful at building fruitful research collaborations.

To make matters worse, Joanie has developed chronic, severe stomach pains. They frequently come on right before she is scheduled to teach and are sometimes so bad she is forced to cancel class. She must watch what she eats very carefully and always be aware of her surroundings in case a physical crisis should hit.

In short, Joanie is miserable and feels powerless to change the many parts of her circumstances that make her feel that way. She feels she must be working or caring for her mother all the time, and that the only time she can take a break is when she is too sick to work. She hates to admit it, but she sometimes daydreams of the stomach pains being so bad that she might be admitted to the hospital and at least have a few days' rest. But she dares not tell anyone about this fantasy, and feels there is nobody she can reach out to for help. She is close to quitting her job and looking for something in the private sector.

Clearly, there are some problems with Joanie's situation that require outside intervention; the misogyny in the department represents a special case for which Joanie may need additional support and mentorship. It might even be that leaving the department would be the right choice, given these problems.

But we think it is also possible that taking the time to be curious about her own situation would be fruitful for Joanie, regardless of what choice she makes. For example, taking some time – even when it seems she has none – for stillness would remind Joanie that she has some control over the course her life is taking. It would recenter her and allow her to get in touch with her own strengths and priorities. It would also enable her to locate where in her body she most experiences both situational and ongoing stress. It could be that her stomach pains are a result of stress, or something more medically serious. Spending some time to interrogate those physical feelings might yield useful information that would allow Joanie to act more meaningfully on her own behalf. Rather than stuffing down the pain, or trying to manage it, Joanie could experience it, breathe into it, and try to figure out what is at its core.

Joanie also has the opportunity to interrogate what kinds of "stories" she tells herself about her professional success. As Beck suggests, Joanie may be living a painful, but ultimately unnecessary story. Like John in our hypothetical case above, Joanie clearly feels trapped in an anxiety-inducing situation. She believes the members of her department don't care about her success, that she is on her own, and that she doesn't have what it takes. All of these things are *potentially* true, but perhaps they are not.

But Beck's questions ask Joanie to do the work on those beliefs anyway. It is possible, for example, that there is a friendly member of the department who might be a good sounding board, or who could suggest resources for Joanie. It is possible that there are women in other departments who might be able to support Joanie in ways her own department members cannot. Perhaps, there are even campus liaisons – administrators or staff who work on gender equity or faculty mentoring issues – who could assist Joanie with defining and addressing the concerns she has. But Joanie will only be able to see these resources and act on them if she begins to see herself as a resourceful person who can help herself, which is certainly a less painful story to believe about yourself than believing you are at the mercy of "sadistic" colleagues. And if your colleagues *are* sadistic (which could happen), then it would be better to leave such a department feeling resourceful and grounded rather than victimized.

The point of all of this is to get in tune with what makes you tick. We all have drivers of our behavior and ambitions. We may, for example, have the ambition to be a leading expert in a certain field of research. This may appear to be a straightforward ambition, but why exactly do you have this ambition? Do we want to be a leading expert because we really want to solve a certain scientific question? (We called this earlier the "itch to know.") Or do we want to be that expert because of the status and job opportunities it confers? Do we want to please somebody, perhaps a parent, former teacher, spouse, perhaps somebody who is not even alive anymore? A seemingly straightforward desire, to be an expert in a certain research field, can be an expression of a variety of underlying motives. Ultimately, such questions boil down to the question, what is it that makes you run?

Do we want to please somebody, perhaps somebody who is not even alive anymore?

The process of embracing a new internal driver takes time and determination. In his book *The Witch of Portobello*, Paulo Coelho (2008) writes: "Reprogram yourself every minute of each day with thoughts that make you grow." Indeed, changing our internal drivers literally involves a reprogramming of the mind. It can be helpful, sometimes even necessary, to have the assistance of a professional in doing such reprogramming. This allows us to be aware of the story that we are telling ourselves and is the first step toward replacing the stories that don't support us by new thoughts that help us grow.

CURIOSITY AS ANTIDOTE FOR ISOLATION

Of course, for curiosity to be effective as a tool for scientists, it should be not just inner-directed but also outer-directed. Taking the time to show interest in others, in fact, is probably the single most important step academic scientists could take to improve their relationships at work and at home. Asking others questions about themselves, and being truly curious about the answers, makes us more connected to others. It shows we care, it takes us outside of ourselves, and it highlights the ways we are in interdependent relationships with those around us. It's also an excellent way to address conflict and to ensure a more positive outcome when there are conflicts.

For example, in an academic environment, we could practice being more curious about our students. When Jen teaches a course on qualitative methods to graduate students, she encourages them to always keep in mind two mantras: "Tell me more," and, "Be willing to be surprised." When students are interviewing study subjects, asking the simple question "Tell me more" is often a wonderful way to encourage others to open up; it also reminds the researcher to not make assumptions about what they've heard, or to fill in the blanks with their own understanding. We can do the same with our students. If a student gives us a perplexing answer in class or an advisee makes a confusing comment in a meeting, we can ask, "Tell me more."

We should also be willing to be surprised by their answers. What we had assumed about students might not be true. For example, we could have some seemingly resistant or even surly students in our classes. It would be easy to assume this is because they don't like us, or the way we teach, or the class. But it is entirely possible there is something else going on. For example, perhaps some of these students are very tired because they have a night job to pay for their education. It is tempting to make all sorts of assumptions about the people we work with, but these assumptions might be incorrect.

Take a moment and ask yourself: how do you see your students?

- Are they simply empty vessels that need to be filled with your knowledge?
- Do they want to be filled with that knowledge, or do they resist?
- Are students loathe to put in the work needed to learn?

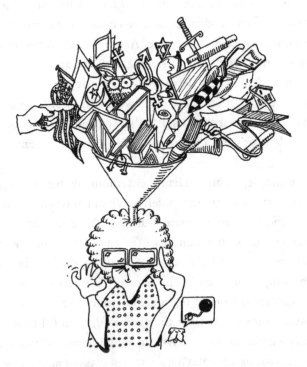

Students as vessels that need to be filled with knowledge.

- Are students becoming less capable every year?
- Do you know students personally, or are they just names on a page, or numbers in a spreadsheet?

Viewing our students as empty receptacles who should just do as we say with a positive and eager attitude can lead to a good amount of resentment and bitterness about teaching and learning, on both sides. In this view of teaching, teaching is indeed a *load* to be shouldered.

By contrast, think about this next set of questions, and how these questions shift your relationship to your students, and to the learning process.

- Are students intrinsically curious?
- If they have lost that curiosity, how can we as teachers help them rekindle it?
- Do we know the names of our students? What else can you learn about them?
- Do we know, at least for some of our students, what drives them, what interests them, and what challenges they may face?
- Do we reach out to our students in ways that go beyond just communicating disciplinary skills and knowledge?
- Do we create an atmosphere of open questioning, or community, or opportunity for exploration in our classrooms?

The first set of questions, when answered affirmatively, expresses a skeptical point of view and positions students as adversaries. The second set positions us to be truly curious about our students, to create relationships with them as *part* of the learning process, and allows us to bring our personality to teaching in appropriate and productive ways. We believe that when we demonstrate curiosity about our students it is likely to spark a mutual curiosity within them, a curiosity in their teacher and hopefully in the topic. When that personal connection is forged, the teaching is not only more effective; it also tends to go beyond the pure disciplinary aspects of our work.

Exercise: how do you see the students you work with?

As an academic you may work with students, whether graduate students or undergraduate students. In this exercise we ask that you write down how you see your students, generally speaking. Do you mostly see the gaps in their skills or knowledge? Or do you focus on their potential? Are you aware of hurdles that they face in their professional and personal development? If so, write down a few of such barriers. Is it your job to help students overcome their barriers? If so, what steps do you take to help students jump over their hurdles?

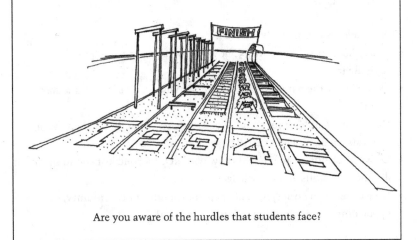

Are you aware of the hurdles that students face?

Another area in which curiosity can be incredibly useful is in our interactions with the colleagues and staff we work with on a daily basis. For staff in particular, it is good to remember that they are often paid less, receive less recognition, and are frequently ill-treated by harried or grumpy professors, administrators, and students. When everything goes well, we tend to take their contributions for granted, but when something goes wrong, we are ready to complain. Worse yet, we often treat them as invisible; we arrive at the office and without so much as a "hello" demand they make copies or straighten out a problem with the budget. But the support staff is essential to the well-being of a university, and the people providing the support have equally rich and interesting personalities as the faculty. So be curious about those that support you! As a first step, you could learn the names of the people that clean your office, take care of your computers or laboratory, or that help you with finances. You might be surprised what happens when you ask somebody, "How are you really doing?", when that question is expressed in a way that communicates a genuine personal interest and curiosity. Such a curiosity in others not only opens doors; it also opens hearts. And when that happens in a professional environment, the workplace changes.

BE WILLING TO BE SURPRISED

For curiosity to work, we have to be willing to question the stories we tell about ourselves, and to be surprised by the possible alternative beliefs that could make our lives better. The beauty of these exercises is they can help us see that our internal drivers are not cast in stone; we can change them. In practice this may not be easy, because our drivers have become habitual, and often we are not really aware of what our drivers are. A first step toward such change is to be aware of our internal drivers, or the stories we tell about our lives, and for this to happen we must be curious in ourselves. The second step is to ask ourselves if we really want to change our behavior.

Here is one final example of an underlying belief that could shape behavior, and how that belief could be replaced by an alternative. Suppose we hold a deep belief that our worth is tied to how much money we bring to the university, along with the number of publications we produce. Such a belief might prompt us as scientists to focus excessively on generating as many research grants as possible and to publish at the highest possible pace. We have to work extra hard to get as much as we can from the world, and spend our lives proving ourselves. This belief structure suggests that success lies on the other side of relentless long hours and accolades, and that if we just get enough of those, someday, we will feel happy.

But what if we replaced that belief with the belief, "If I approach my work and home life from a place of passion and curiosity, new resources and opportunities will make themselves available?" Hard work is still very possible and probably required under this belief system, but we are motivated from a place of intrinsic satisfaction rather than external reward. One of our interviewees put it this way:

> We ask a lot from young academics who start a career as tenure tracker. In five years they should demonstrate their ability to attract financial support for their work, publish and become known in their field (usually "evidenced" by the H-index), to be a gifted teacher and obtain didactic qualifications, to build and supervise their own group ... Several tenure trackers feel they compete with each other ... others feel they need to work on their H-index.
>
> I try to explain that they should look at it differently. Start by sharing your results in peer reviewed journal papers and talk about your work in conferences, workshops, and other meetings. Colleagues will start to know you and your work. You can get to know them and you can interact with them, the H-index is a consequence, not a goal ... The ones who are able to focus on what they like about their jobs, enjoy doing research, teaching, and communicating about research results often are successful.

In research, therefore, it also helps to be curious about others. In practice, research often is a collaborative effort. Many of our ideas are sparked in conversations with others. Often our research questions can be addressed by insights or data from others. And working with others can greatly improve the enthusiasm we have for our work. In most cases, research is most effective and most rewarding when it is a collaborative activity rather than an individual effort.

In practice, research often is a collaborative effort.

In sum, it is worthwhile to have an understanding of the deepest drivers of our behavior, because these drivers manifest time and time again in our life. For example, if we feel that the world ignores us, we likely behave as the scientist who has an aggressive and dominant stance at scientific conferences because we feel we need to behave in this way in order to be heard. In practice, this behavior often is counter-productive because it is likely to generate resistance in our colleagues. On the other hand, if we are truly curious in the world around us, we likely spark interesting ideas, insights, and collaborations while being at a scientific meeting. This attitude will probably draw colleagues to us. So two researchers at the same conference can behave completely differently because their internal drivers are different, and the response from their colleagues is likely to be different as well.

In fact, in the process of getting insight into our behavior and possibly changing that behavior, our interactions with other people can be very important. In practice, we often are nearsighted about our own shortcomings. For example, when we write, it is very difficult to spot the errors we make in our writing. This is why it is so useful to ask somebody else to proofread our writing before we submit our manuscripts or proposals. It is for the same reason that writing with co-authors is so effective. What holds for writing also holds for our behavior; those around us can provide valuable feedback on how we behave; they have a "freshness of mind" concerning our behavior that we lack. So we want to encourage you to build up a network of trusted friends and colleagues who are curious in each other, who challenge your behavior and provide new avenues for thinking about the way you do things.

The absence of such feedback in academia may be one of the reasons why scientists display so much exotic and aberrant behavior. For an amusing tour of deviant behavior of academics, we refer the reader to the *Field Guide of Experts* by Oxman, Chalmers, and Liberati (2004). Most universities have a system for evaluating faculty, but in practice the evaluations focus on quantitative measures such as the number of publications, the number of classes taught, and the number of students supervised. But how often do we get corrected in our behavior? And as we grow though the academic ranks and become more senior, the corrections become even more rare. It is no wonder that the caricature of the "nutty professor" is so popular.

5 Listening

> I hope that everybody has some activity that gives their body so much joy that it can tell the mind to shut up. And then the mind finds itself on vacation, without the perceived need to output a lot of silly chatter. And lo and behold, when the output spigot is turned off, uncensored input flows in.

John Turk, *The Raven's Gift* (2009)

We talk a lot. And when we're not talking, there is still a lot of "noise" around us – whether in the form of television, texts, or social media posts. In fact, we live in a society where talking and noise are such a big part of everyday life that many of us find silence difficult to bear, especially when we are in the company of others. Our tendency is usually to fill silence with conversation, to override our discomfort with silence by filling the void. Contrary to our normal communication habits, when we are silent, present, and consciously tuning in to the words of others, we communicate that "I am listening." This willingness and ability to listen is usually much appreciated and it gives others permission to open up.

Listening is an art, and this art is the topic of this chapter. The word "listening" is used here in a broader sense than just taking in auditory input. We use this phrase for tuning in to information, spoken and unspoken, from both ourselves and others. We also define listening in an active, rather than passive, sense. It is a conscious choice to engage in communication with others and ourselves in a slower and more deliberate way than we might normally be used to.

WHAT DO YOU LISTEN TO?

In these times, we often get to choose what kinds of information come our way – we have control over which people we talk to, the material that we read, and which television shows or movies we watch. We may not be able to control the content in those sources of information, or we may disagree with the values that are communicated in them, but we have some choice over which of these physical channels of information enters our day-to-day existence. We might feel, even, that we are pretty discerning about which voices or opinions we allow to influence us.

But this chapter aims to identify voices in our lives that are less tangible, and frankly, sometimes easier to ignore – although we do so at our own peril. These subtler voices include the expectations of those around us, the mores of our workplace and home lives, the

We are continuously telling ourselves a story about what happens to us, often this story is rife with judgments.

voices of our students and colleagues, and perhaps most significant, the conversations that take place within. Just as you make choices about which physical channels of information you take in, we believe you also make choices about which of these implicit voices you listen to, and how.

Earlier we mentioned the voice within; this brings us to the concept of the *internal dialogue*. Whether we are aware of this or not, we are continuously telling ourselves a story about what happens to us. Often this story is rife with judgments. You may recognize the following example: you have an issue to resolve with somebody else. Before meeting that person you may think about what you are going to say. You might even "rehearse" words that you would never say in reality. In the process, your conviction of how you might have been wronged by the other person becomes deeper and deeper. This internal conversation may darken the meetings you have with that person, and the process poisons not only your personal interactions but also your own mind. This inner dialogue can be so filled with criticism of others and of ourselves that one speaks of the *inner critic*. On the other hand, if you can admit from the outset that your understanding of the situation may be partial, if you can allow for the fact that you may not know the whole story, if you can assume that the other person acts in good faith, you are likely to have an open conversation that leads to a constructive outcome. In this example, two different mindsets in the same situation not only produce completely different perceptions of the situation but also lead to completely different outcomes.

To make matters more complicated, we don't carry out the internal dialogue alone; we do it collectively with others. In fact, many of the beliefs that we have are collective beliefs. To name an example from the introduction, there is a common belief among scientists that our work is "never enough." We therefore carry out our internal dialogue in conjunction with others. Since many of the collective beliefs are rarely explicitly talked about, we may not even be aware how our thinking is influenced by others or by the local culture of our time. This raises the important question: who sets the tone for the internal dialogue?

For example, especially when we are in those fraught years before tenure, or if we are in a department that heavily emphasizes competition and accomplishment, the dominant voices in our lives might be those that emphasize professional achievement above all else. These voices might tell us that we need to get the next promotion, be a part of important committees, generate much research funding, be praised publicly, and often, be the best. These outward pressures relay the collective beliefs of our research environment. We are not arguing that these voices are wrong or should be ignored. Sometimes competition is healthy, and striving to achieve can help us realize some of our career goals. Our point is that, when not balanced by a broader perspective, this "achievement voice" might easily lead us to the belief that things are "never enough," or that *we* are never enough, a belief that we discussed in Chapter 1 on Harmony. Sometimes the achievement voice yells loudest at us, or seems most important, and we would suggest that when this seems to be the case, it is time to make space for other voices.

Another voice comes from the "itch to know" and to our sense of wonder, which we discussed in Chapter 4 on Curiosity. Listening to this voice is important in curiosity-driven research, and it can be a powerful and effective tool, akin to an academic intuition. At the same time, when we exclusively listen to this voice, we might turn into that out-of-touch scientist who is oblivious of the environment in which she operates. As is the case with the achievement voice, we can use the voice of curiosity to our professional benefit, but we should be watchful that it doesn't drown out other perspectives or messages.

While the achievement voice is often an external voice (one we have perhaps internalized!), the curiosity voice is an internal one. There are other internal voices that are important, including the intuitive voice. Along with the external channels of communication we discussed above, we are bombarded with information from outside. We can pick up a phone or use our computer and instantly speak to people on the other side of the world. Emails and social media come to us at an increasing pace. Our brains have become wired to these stimuli over time, and most of us who are professors have noticed that we spend increasing quantities of time responding to these messages. As a result the temptation may exist to pay exclusive attention to these external voices. But what about the internal voice, our intuition? The intuitive voice doesn't come with a notification sound on our smartphones, so it requires a different kind of tuning in and paying attention. The intuition – often called "the still small voice within" – can be a powerful guide that provides us insights, wisdom, and sometimes practical guidance. But this only happens when we pay attention. Do you take time to be silent, to be away from the external voices, so that you can hear your internal voice?

One may think of intuition as an expression of the right brain, which, as stated earlier, we use as a metaphor instead of a neurological entity. If you spend most of your time engaged in left-brained activities, as scientists and engineers are trained to do, your right-brain "muscle" may be underdeveloped and, paradoxically, your left-brained self may suffer as a result. The left and the right brain have different stories to tell, and listening to both halves of the brain can be useful, not just because of the variety of insights and experience that is gleaned, but also because of the balance that it brings. In the modern research environment, we don't talk much about the value of the right brain, or the intuitive self. Many of us believe that only rational thinking is important. Yet research increasingly suggests that rational decision-making requires both sides of the brain, and takes place using both sides of the brain, and those who actively engage in activities that trigger both sides of the brain are more effective and creative problem-solvers overall (Lombrozo, 2013).

Experience tells us that we need to feed both halves of the brain if we are to feel harmonious and in "flow." What does it mean to "feed" the brain? This phrase refers to making conscious choices regarding what we listen to. Reading a scientific paper most likely feeds the rational self. Listening to free-flowing music with many minor chords feeds the intuitive self. These two activities engage different parts of our thinking and feeling selves and can produce different types of cognitive and problem-solving outcomes.

An anecdote illustrates this point. Roel was chairing a committee at his university – and remember, he works at a university devoted to science and engineering, where left-brained work is highly valued. This committee was charged with doing some difficult administrative work but also creative work, work that required they evaluate an existing program and imagine alternative futures for it. Given the committee's charge, Roel decided to approach committee chairing differently from usual. Instead of just jumping into the committee's meetings, he started each meeting with some music, a quote, or poetry. This sharing took little time, but it changed the meetings because it made the group tune in to their right brains before diving into the problem-solving, vision-setting aspects of the meetings.

Starting a meeting at a university in this way was scary at first, and it made Roel feel vulnerable. We acknowledge that there are plenty of committees and audiences for whom this kind of experimentation might backfire – you must carefully read your audience before trying something like this. But Roel read this particular group correctly and found that taking these creative risks had unexpected and beneficial side effects. Even though the meetings were dealing with a contentious and potentially difficult topic, everybody on the committee was willing to be more vulnerable because Roel had been, and because the committee activated the right brain, the meetings were probably more harmonious than they would have been otherwise. The committee, which was also charged with gathering input from various stakeholders on campus, was also better able to listen and stay open to what they heard. Other committee members eventually joined in and brought their own creative contributions to begin the meetings with. Maybe this experiment worked so well because it helped the members of the committee to tune in to the present in a way that committee work rarely inspires. It reminds us both that there is a world beyond the task at hand and that the task at hand can be approached in multiple ways.

Exercise: tell me more

Here is an interesting exercise you can do with a friend. Sit in front of each other and take turns sharing something important that happened to you recently. When you are the listener in the dyad, the only thing you can say is "tell me more." Otherwise you stay silent. Do this for just a couple of minutes, and you are likely to be amazed what you will learn from the other person in such a short amount of time. Why? For most of us it is easier to share personal issues, commiserate, or give advice than to bear the silence.

PAY ATTENTION AND CHOOSE WISELY

We only listen and learn when we pay attention (Medina, 2008). The act of paying attention is needed to really take things in. We can all hear things without tuning in to them – almost everyone has had the experience of talking to somebody, nodding, and responding, even while their mind drifts off to something else. Even though you could hear every word that was spoken, you end the conversation confused or with a sense that you didn't really connect. This can be an alienating and frustrating experience.

In other words, listening is more than letting your eardrums vibrate to somebody else's voice; one needs to be attuned and attentive to take things in. But it also works the other way around. The issues that we pay attention to tend to grow in our life. For example, when we listen to gossip about colleagues we may be tempted to gossip ourselves, or to believe that others are gossiping about us. This results

What we pay attention to tends to grow.

in us feeling paranoid or uneasy: again, alienation is the result. On the other hand, when we focus on the positive contributions that our colleagues make, we tend to appreciate them more, and because of this appreciation they are likely to work harder to make positive contributions. This also makes it easier to appreciate ourselves and our own efforts, and to step outside of that "never good enough" pattern so many of us fall into. The bottom line is that what we focus on grows, and what we choose to listen to will increasingly capture our attention.

A story from social science research illustrates this effect. A prominent communication researcher writing in the 1970s, George Gerbner, developed with colleagues what he called "cultivation theory" to describe the effects of media exposure on consumers (Gerbner et al., 2002). They argued that the more we watch television, the more we begin to believe that what we see on television reflects the actual world – media can "cultivate" our ideas about the world. This research then led Gerbner to develop the "mean world syndrome" hypothesis, which says that heavy watchers of TV tend to believe the world is much scarier and more violent than it really is. Heavy TV watchers who spend much of their time viewing the evening news and crime dramas may develop ideas about the world that don't match up with reality.

We are suggesting that the same might be true for academics. Like those who watch too much television and believe the world is primarily a mean place, academics who only pay attention to their "achievement" or intellectual curiosity voices, and make little space for voices of intuition or creativity, may see the world primarily as one of cutthroat competitiveness. Shifting our attention to other voices, however, may allow us to do different kinds of work and to feel more at peace and harmonious in our day-to-day lives.

Our ability to choose, at least to a certain extent, what we listen to offers us tremendous freedom. By tuning in to the voices that make us grow, we actually do grow. Do you remember the words of Paolo Coelho we quoted in Chapter 4 on Curiosity, in which he invites us to reprogram ourselves with thoughts that make us grow? Carefully choosing the channels that we listen to is an important aspect of choosing thoughts that make us grow. In this chapter we discuss the art of effective listening, of opening up, and the different voices that we listen to. But before we continue, we first invite you to the following exercise.

Exercise: record what you listen to

What is it that you listen to? This question may be difficult for you to answer. Perhaps you know the answer right away, or perhaps you think you know the answer. In reality, we often are only partly aware of what we listen to. In this exercise, we will take a scientific approach. Instead of speculating, we collect data. Here is the exercise:

> For 24 hours, write down what you listen to. As described above, interpret the word "listening" in a broad sense. You might listen to a lecture, to a friend that you talk to, to the television news, to a novel that you read, to your internal dialogue during a faculty meeting, to the thoughts that come up when you wake up, and perhaps to your dreams at night. Write it all down, together with an estimate of the time you have devoted to each voice.

We suggest that you approach this exercise as a scientific data-collection project that has the aim of making an inventory of the voices that speak to you. That said, it is just an exercise, so don't drive yourself crazy trying to get an exact tally of your every thought or media exposure – make your best guesses and fill in where you can.

Notice the following: which voices feel supportive? Which do not? A supporting voice lifts us up, inspires us, stimulates our creativity, and helps us reprogram ourselves to be a better person, to feel better. In contrast, the voices that don't support us drag us down, worsen our mood, make us somber and dependent, anxious or judgmental, and trigger thoughts and behavior that don't make us a better person. Now move on to the next step:

> Take the list of voices that you listened to and decide for each voice whether it supports you, or whether it does not support you. Take a piece of paper with a column for both types of voices, and place each voice in the appropriate column along with the time you have spent listening to that voice.

Once you have done this, add the times you have spent listening to the voices that support you, and to the voices that don't support you.

What have you learned from this exercise? Do you spend most of the time paying attention to the voices that support you, or do you listen more to the voices that don't support you? Likely, you will listen to both. But we have a choice in what we listen to. Perhaps we can tune out the voices that don't support us, and focus more on the voices that do support us.

Take a moment to reflect in your journal or workbook on how you might build more time into your day for supportive voices, and how you might marginalize those that don't support you. Think in terms of small changes if that's helpful.

WHAT DOES IT TAKE TO LISTEN?

Okay, now we are going to shift gears a little bit. In the last section, we focused on how you might pay attention to the kinds of voices you expose yourself to in order to better construct a supportive environment for yourself. We talked about priming your metaphorical "right-brain" more frequently to enable you to do better, more creative work, and about how to notice when you are exposing yourself to too many "mean-world" kind of voices.

In this section, however, we are going to focus on how you yourself can become a more effective listener in terms of your professional relationships with others. This is not about controlling which channels we are exposed to, like we discussed above, but how to become a more effective and compassionate listener *for others*.

So, what do we mean by "listening" in this context? For most of us it is difficult to really listen to others because we are so accustomed to talk incessantly. We academics are probably especially programmed to use our own voices, dominate conversations, and win arguments – if you're anything like us, you find it hard to help yourself from doing this when communicating with others. The same may be true for our internal dialogue, which can be relentless in its own right. Of course, the internal dialogue is totally natural! For example, you might be driving and you probably continuously narrate in your head what is happening: "Watch out, the red car may overtake the truck, watch for the indicator, oh no, he is moving left without the indicator, slow down, hey, isn't that my friend Joan behind the wheel?" Or you replay the argument you had with your spouse or supervisor while driving, coming up with a better comeback for next time she has you cornered. And so it goes on and on and on. The internal dialogue acts as a filter that colors how we see the world, and how we listen to it. This filtering might be accurate, but it could also blur or distort the way we perceive.

At a more mundane level you may have noticed this in conversations. You are talking to somebody else, and at one point you hear something and you feel that you must react. In fact, you are dying to interject something! While the other person is still speaking you are ready to formulate your comment, and you may actually be so preoccupied with figuring out what you want to say that you don't listen at all anymore. As a result you might miss that final piece of information that puts the earlier statement of the other person in a different perspective. This happens to Jen all the time, still – she gets so enthusiastic about what the other person is saying that she can barely contain her reaction. But talking over others has rarely led to good communication and does not convey to the other person that their input is valued. In an academic setting this often happens in faculty meetings.

We should note that there are cases in which staying too much in the role of listener can prove very difficult or even counterproductive. Some academic personalities are very strong and, in some cases, may prove abusive or intractable. If you are interacting with someone who herself is not good at listening, or with someone who dominates every conversation he enters into, then conscious interruption may be necessary. In this chapter, we are not talking about situations in which there are repeated abuses of power where certain voices dominate to the exclusion of other perspectives: those situations require different interventions than the ones we describe here. Instead, our focus in this chapter is on those who find they are struggling with connecting or communicating with others and suspect their inability to listen may be at fault. We are writing for the self-reflective reader who wishes to improve the way he connects with others through improved listening.

For that reader, we would say that the upshot is this: to really listen one needs to be quiet. Being quiet means both to not speak physically *and* to really give yourself over to what you are hearing. It can feel impossible to do this without filtering or a constant stream of judgments that normally accompany our interactions with others. One way to get to a more uncensored form of listening is to repeat back to another what you think you have heard them say, offering them an opportunity to correct you or expand. This encourages you to step out of reaction and into interaction and is a helpful way to send your more judgmental chatter on vacation.

One of our colleagues shared this gratifying reflection with us:

I just turned 44 and celebrated by going camping with my family. I had been working very hard for several weeks, putting together a draft of my tenure dossier – finally, nearly ten years after receiving my Ph.D. Around the campfire, we followed our family tradition of going around and saying what we love about the person who is having a birthday. My younger daughter said that she appreciated that I really take time to be with her, unlike other dads she knows about.

This means so much to me, because I have reminded myself over and over to be intentional about taking the time to just be with my kids. I don't focus on special outings or fancy vacations – although we have enjoyed some of those along the way. Instead, when one of my daughters interrupts me to check in about something, I try very hard to consider whether I can take a short break and really give her my attention. I often can, even though my instinct is to explain that what I am doing is really important. So, hearing that affirmation, and feeling loved and appreciated by my family even as I am hurtling toward tenure review, means the world.

This is an example of a busy academic and parent who made a conscious decision to listen not only to his professional desires but to his personal desires – and to his daughter. This allows him to have a sense of peace and harmony as he "navigates the uncertainty" of awaiting the year-long tenure decision.

In Chapter 1 on Harmony, we discussed the concept of the outer world and the inner world. In a similar way, each of us can be seen as an outer person and an inner person. The outer person is the body with all its manifestations such as the words that we speak. The inner person relates to the true personality, the inner thoughts, and if you appreciate the concept, the soul. When meeting somebody superficially, we focus on the outer person. We see her body, watch it move, and listen to the words that she speaks. At this level it is as if we watch a movie with events in the external world.

But what does it take to perceive the inner person, the inner man or the inner woman? John Turk has provided the key in the quote at the start of this chapter: we need to be quiet and still our mind. By truly listening to the other without telling ourselves what he or she is telling, we can truly listen. By carefully listening or seeing without filters or preconceptions we can often observe telltale signs of the movement of the inner person. Describing the sense of seeing instead of hearing, Laurens van de Post writes in his book *The Heart of the Hunter* (1961) that "vision [is] complete only if we [see] reality with both the outer and inner eye ... I [believe] one [does] not know human beings really until one [sees] them that way as well." He speaks about the outer eye and the inner eye, just as one can listen to words in a literal sense, or to what is said between the lines. For some of us the ability to go beyond appearances comes naturally, for others it is an unknown faculty, and for most of us it is something we can cultivate. How do we cultivate this ability? We can do this by being silent, and by careful hearing and listening without filtering. This takes training and dedication, but it can be done. The bottom line is, if we really want to listen we need to stop talking.

LISTEN TO YOUR SELF

We talked above about hearing or seeing the inner person in somebody else. When others have an outer person and inner person, then so do we. Who is that inner person? What comes up from deep in your personality? What are the dreams and desires that most deeply live in you? What is the story that your inner person has to tell? According to psychotherapist and former monk Thomas Moore, our inner person has an agenda. Perhaps we are aware of this agenda, but perhaps we are not. Moore makes the point that the inner person is a powerful driver, whether we are aware of it or not, and whether we like it or not. And he states that the drive of the inner person can be so strong that when we neglect that voice, the inner person is likely to assert itself anyhow. So it is important to pay attention, to listen to the inner voice, and to find out what our deepest desires really are.

If we truly express ourselves in our work, then our work reflects the deeper aspects of who we are. Moore (1992, p. 186) writes "our work takes on narcissistic qualities when it does not serve well as a reflection of self. When that inherent reflection is lost, we become more concerned instead with how our work reflects on our reputations." When our work is disconnected from our deeper self, it does not serve the inner person – this is what we mean by finding the good work in our professional lives. If our work does not serve the inner person, then who or what does it serve? It becomes an activity that serves itself and that can result in feelings of frustration and alienation. In an academic context such work may be focused on cranking out publications, or on becoming famous for the sake of being famous. These activities, when overdone, can easily lead to a narcissistic self-glorification rather than contributing to the world by doing good work, however we have defined it when we set a vision for our lives (see Chapter 3). As Moore states, when we only focus on work for work's sake we care more how our work is reflected on our reputations, than on how our work reflects our inner self.

We don't mean to say that our reputation is not important. It is important because our reputation can support new opportunities such as our finding research grants, collaborations, and job offers. But an excessive focus on reputation can cut us off from our vision for our life and our dreams and desires. Listening to that inner self can prevent such a disconnect.

LISTENING AND SPEAKING TO OTHERS

Above we discussed the importance of listening to the voices within, but that does, of course, not mean that we don't pay attention to the voices of others. To communicate effectively with others can be both an art and a challenge. For example, many of us are shy. Even the most outspoken of academics may have introvert tendencies or social anxiety. Jen, who has never been accused of introversion, really struggles with social anxiety in academic settings and has to work hard to focus her attention on others and not on herself in order to connect. Such shyness or awkwardness may make us reserved and reluctant to reveal anything personal. This might lead you to think that your colleagues are "just" interested in their science, or that they have flat personalities.

Communicating personal issues can be challenging for scientists because such issues go beyond the "objective" standards of science. This is especially the case for conversations among men, because for many males the appearance of being sensitive and emotional is perceived as a sign of weakness. So as a result of our reluctance to really share openly, we often communicate between the lines, and allow things to go unsaid. This can lead to misinterpretations or a lack of connection. As a speaker, we might not say what we mean because we want to test out the waters – to gauge where someone is at, or if we can trust them. If I am at lunch with a new colleague and they ask how I am doing at work, I might say, "I really like the work I am doing, but . . . ." Then the conversation is at a crossroads – our colleague might respond by interpreting our trailing off as a desire not to share more. She might quickly change the subject and start talking about her new puppy. Or, another choice might be that she puts down her fork, takes a breath, and repeats, "You really like the work you are doing, but . . . ." And then

For many males the appearance of being sensitive and emotional is perceived as a sign of weakness.

allowing for some silence. This gives us a chance to figure out whether we want to share more and deepen our connection.

Another of our colleagues described an opportunity he had to listen to a cry for help from a student, which permitted him to intervene meaningfully and compassionately on the student's behalf:

> When I first arrived as department head there was a student who would enroll one semester and get excellent grades, then enroll the next semester, attend class for a while, then disappear, resulting in automatic Fs in all his classes. He repeated this pattern a couple times and was about to do it again when I arrived on the scene.
>
> My faculty asked me to try to connect with him and see if there was anything I could do to break this pattern. This particular student was depressed as a result of stressful personal circumstances including extraordinary pressure he was feeling from one of his parents. In evaluating his transcript and seeing the disastrous impact on his GPA of so many automatic Fs, I realized this student had dug a hole for himself that would take forever to remedy.

Being new at my job as department head, and not really understanding what was possible or impossible, I took the student to a senior administrator, told his story, and explained the situation. The senior administrator, persuaded in the moment, went to his desk, logged onto his computer, and expunged all these automatic Fs from the student's transcript. I was thrilled at this outcome. It gave the student a new outlook on life and freed him to move in a new direction without a huge burden of academic failure.

In retrospect I privately marvel at what must have been an "illegal" act on the part of the administrator to whom I made the appeal.

In this story, our colleague first took the opportunity to listen to his colleagues, who asked him to intervene; then he truly listened to the student, and recognized a destructive pattern that could be remedied. He likely changed this student's life for the better, when it would have been just as easy to avoid listening, to write the student off as a failure.

As we indicated earlier, real listening is difficult, but it takes even more attention when the message is delivered implicitly. We would say this: academic and graduate advisors, and mentors too, listen up! Our students and even junior colleagues often lack the language to express their worries, concerns, or desires. They may be trying to open up avenues of communication with you but they don't know how to do it and are only able to offer these little openings in conversation. They need help in opening up. You could provide such help by listening carefully for telltale pauses, or you may need to pay attention to their body language. You could try using open-ended questions and avoid filling in their sentences so that you have the opportunity to learn from or be surprised by what they are going to say. In almost any situation, repeating a statement can help open the door to communication. And sometimes the only thing to do is to be silent and to be available. Such listening actually comes naturally when we release our own desire to speak. And when we do that with a genuine interest in others, we open a door to a deeper level of sharing.

Students may be trying to open up avenues of communication with you.

This leads us to what is really the crux of this chapter: like curiosity, listening is an important tool in fighting the sense of alienation many of us feel in academe – in fact, listening is curiosity's counterpart! We can ask questions, but then we must also listen for the answers. These are such important tools because they enable us to connect meaningfully with others, and connection brings lasting meaning to our lives. Listening, and being effectively heard ourselves, allows us to feel that we belong and that we matter. We want this for ourselves and our loved ones, and we want to be able to offer it to others.

LISTEN TO YOUR BODY

Our bodies are amazing. The living organism that is "us" operates in incredibly complicated ways. It is self-renewing, and most of the time self-healing. It operates for a long time, usually significantly longer than half a century. What is even more amazing is that we can neglect the needs of the body for a long time, and still it works.

We academics neglect our bodies in so many ways: we deprive them of sleep. We sit them behind a computer or lab equipment all day and don't move them. We over-caffeinate them or feed them junk or subject them to highly stressful situations. We isolate them from human touch and affection. We force them back to work when they are sick and tired. And sometimes, we get away with doing all of this, even for a long time. But there is a limit to the degree to which we can neglect or abuse the needs of our body, although it might take a long time before this becomes manifest. Our bodies are indeed amazing!

Still, we feel strongly that our bodies function as a key barometer of the levels of harmony in our lives. When something is out of whack, we believe that it is pretty likely you'll see signs of that imbalance in your physical health. Perhaps you feel recurring or chronic pain or discomfort. You might sleep enough, but perhaps your nights are filled with restlessness. Do you have medical problems, such as infections, that recur often? Sometimes these signals from your body have a story to tell.

We believe this so strongly because we have experienced it. It wasn't until writing this book together that we realized we have experienced parallel health challenges related to overwork in our early careers. When Roel was in his twenties, he had regular back pain. He visited many physical therapists who provided short-term physical relief, but the back problems would return. Then he visited a physical therapist that said on his first visit, "If you don't learn how to heal yourself, you will have back problems for the rest of your life." She refused to touch him and taught Roel to listen to the story that the back was telling him. After a while the story became clear, because his back pain would start when he was taking on too much and was pushing himself too hard. His back actually told Roel at those times to take his foot off the accelerator and go a little slower.

Amazingly, Jen had the *exact* same experience early in her career. She would spend days in bed with debilitating back pain during her first years on the tenure track, at a time when she felt she needed to be most productive (and when she had two children under the age of three). Jen usually addresses problems in her life through *planning* and *action* (hello, Type A!), so she tried everything to heal herself – yoga, massage, chiropractic adjustments, muscle relaxers (she hated those), and exercise (a well-meaning colleague suggested Jen needed to work on her "core," not exactly what a new mom needs to hear, and counter-productive in this case).

None of it worked. Jen remembers lying in bed in agony, crying and wondering if she would always feel like this. Finally, a family friend gently suggested that back pain is usually our body's way of communicating some other need to us, usually the need to rest more. That friend also pointed out how frequently Jen got sick with respiratory infections, and (again, gently) suggested that Jen was run down and needed to back off the super-mom act. This is probably the worst thing in the world you can tell an overachieving, externally focused human like Jen, and yet it was exactly what she needed to hear. It took several years, but slowly Jen learned about the notion of "self-care" and how to make rest a priority. The back pain went away. It comes back sometimes as a friendly reminder for when Jen is overdoing it, and she is often grateful for that very clear messaging system.

These anecdotal cases do not mean, of course, that this is the story behind every back pain or illness, and it certainly does not mean that every medical challenge should be interpreted in this way. In no way do we mean to blame people who suffer from health concerns for their illnesses. The point we do want to make is this: sometimes we academics are so good at living in our brains that we get very good at overriding the needs of our bodies. In doing so, we are missing out on an incredibly important feed-back system that gives us information about where our lives are and aren't working.

Why would you listen to your body? There are a variety of reasons. For one thing, our body may last longer when we listen to its needs and take good care of it. We might just feel better when our body is in good shape. But even our science, and other aspects of our work, may progress better when our body is well. Research has shown that students learn better when they exercise (Medina, 2008). There is a perfectly logical explanation for this; when we exercise, the flow of blood through the brain increases, and as a result we think better. We find this is true for faculty as well as students – regular exercise, in whatever form that works for you, will make you a better thinker, writer, and colleague.

Thus listening to our bodies, and taking care of our bodies, is not only helpful for being healthier and feeling better; it may also help us to think better. John Turk writes in the quote at the opening of this chapter about activities that give the body so much joy that it can override the mind's incessant chatter, and that it is a way to tap into our intuitive, uncensored powers. The body can thus help us tap into these powers! So the next time you feel de-energized and stuffy and your work does not progress well, instead of thinking that you are "too busy to exercise," you might consider going outside, get some fresh air and get the body going. This may be exactly what the body is telling you when you felt stuffy and de-energized.

So how do you listen to your body? Many of us have become accustomed to treating our body as a piece of equipment that we take for granted. The story is the same as it was before: we need to be still and be aware what we feel. Once we do that, we might discover that we feel really tired, that certain muscles are tight, or that we have a mild headache again. Such symptoms might creep into our life slowly, and because the physical response to our life style develops gradually, we might not be aware of it. Our suggestion is to develop habits that allow you to tune in to how your body feels – meditation practice, yoga, morning stretching or afternoon walks, journaling, regular exercise, playing soccer – all give you time to feel your body and to do so without judgment. Scan your body mentally, listen for persistent patterns in what the body feels, and be open for the message of recurring physical problems.

Our body that has become a piece of equipment that we take for granted.

You have one body. It may serve you well. You owe it your attention.

Exercise: listen to the big picture

In this exercise we will listen for the big picture of your life.
In particular, we listen for harmony or dissonance. We ask that you sit in a quiet place, switch off the internal dialogue as well as you can, and listen for the combination of "tones" in your life. As you do this, think of your work, your social life, your hobbies, your body, your dreams, your sexuality, and your spiritual/inner life. What do you hear? Specifically, do you hear a natural harmony, or do you hear a screeching dissonance? Perhaps you hear both, depending on which parts of your life you are thinking of. Record somehow what you discover. You could make notes, make a sketch or drawing, or record your speech.

Put your notes, or other medium of recording, away for a few days, and then revisit them to search for the message that is contained in what you wrote down. Are there aspects of your life that are particularly congruent with who you are and what you would like to do? Are there aspects that feel unpleasant, out off order, or simply feel "wrong?" Based on what you discover, would you like to make changes? If so, do you know in which directions you would like to change? Write down whatever comes to mind so that you can use your impressions to help you grow the harmony in your life.

6   Compassion

> Beautify your inner dialogue. Beautify your inner world with love
> light and compassion. Life will be beautiful.
>
> –Amit Ray, *Nonviolence: The Transforming Power* (2013)

Compassion is perhaps the most difficult of the seven characteristics
to broach in a book such as this one. More than any of the other traits
we've examined thus far, compassion can mean very different things
to different people – we all have associations with that word (positive
or otherwise) and yet there is not one shared meaning that makes
sense across contexts. Is compassion love? Sympathy? Empathy, the
ability to walk a mile in another person's shoes? Understanding?
Forgiveness? Kindness? Or is it weakness, codependence, or enabling?
Does being compassionate set us up to be fooled, or taken advantage
of? Does compassion deserve a place in the academic environment?
Where do we establish our personal and professional boundaries if we
are also trying to be compassionate? Does a commitment to compassion also require a particular ideological, political, or religious stance?
Are we born with compassion, or can we develop it? And what does
compassion look like in practice?

Frankly, we don't know the answers to all of these questions.
It would probably have been easier not to address the concept of compassion at all – as bound up as it is with notions of love, sensitivity,
forgiveness, boundary-crossing, politics, and religion as it might be.
These can feel like unwieldy or uncomfortable subjects to broach within
the context of the academic workplace. Yet, despite these challenges, we
feel compelled to articulate what we mean by compassion in the professional settings that scientists most often find themselves in, because we
believe that cultivating a compassionate heart is a key ingredient to
becoming a joyful person both in and beyond the workplace.

Before we discuss compassion, we first discuss what compassion *is not.*

- *Agreeing for the sake of agreeing so that we please others or avoid conflict or difficult conversations is not compassion.* Superficially, agreement seems to promote alignment and harmony. But when by agreeing we do violence to ourselves or to our values, when such agreement sets us not in a path of "right action," then we are violating ourselves, our principles, or our peace of mind. Although such agreements momentarily seem to lead us on the path of least resistance, they can express avoidance concealed under a cloak of fake compassion. In the long term such false agreement can easily lead to resentment, stagnation, and discord.
- *Emotional resonance is not compassion.* Do you notice that in conversations with others who are angry, sad, or otherwise emotional, you may have a natural tendency to agree or go along with these feelings? You may talk to a colleague who is angry. How easy it is to become angry as well and to confirm how your colleague has been wronged or mistreated. Or you talk to somebody who is sad, and before you know it, you are wallowing in the same sadness. Roel, being a physicist, calls this phenomenon *emotional resonance*, because our emotions resonate with the person we are talking to. And as with any physical resonance, when we push the resonance the motions only become stronger. And because we strengthen and deepen the emotions of the person we talk to by blindly resonating with these emotions, we only strengthen the anger or deepen the distress in others. We don't do others a favor by such reinforcement, and therefore emotional resonance is not compassion.

So what is compassion? There is a growing body of "compassion research" that suggests, in fact, that compassion can be defined differently according to one's disciplinary or ideological approach. A neurologist will think about and measure compassion differently than a Buddhist monk might, for example. Neuroscientist Tania Singer has done extensive work on compassion, including compiling a fascinating multimedia e-book on the topic, which represents a range of definitions and conceptualizations of compassion, as well as frameworks for "compassion training" (Singer & Bolz, 2013). While it is beyond the scope of our book to provide a typology of these different approaches and disciplines, we build on Singer's edited collection to highlight five "compassion practices" that scientists might use to build compassion into their professional and personal lives.

Briefly put, we believe that compassion is the ability to notice when we or others are suffering, to desire to alleviate that suffering, and then to act to prevent future suffering. So, there are three elements: noticing, desiring, and acting. If we notice but do not desire to change the situation, or act to change it, we are probably living a somewhat shallow, careless, hurried, or over-intellectualized version of our lives. If we do not desire to change or alleviate our suffering or the suffering of others, we are also lacking compassion, and must perform some serious introspective work to understand why we allow our suffering or the suffering of others, and we must understand that this comes at great emotional and personal cost. And finally, compassion is not just a theoretical platitude; it requires that we act in some way in order to change our circumstances or the circumstances of others.

These are the three elements of compassion; but they can be further broken down into or "operationalized" as five practices. Before we describe these five practices – some of which will be familiar from other chapters – we wish to argue for *why* compassion is important for academics in particular:

- Because we deal with so many people, many of whom are often particularly sensitive to our judgments about and interactions with them. We are often put in the position of judging or evaluating others based on their education, their work ethic, their productivity, their "collegiality" or personality, and their intelligence. And we ourselves are often judged on the same. We fool ourselves if we believe that such judgments do not have the potential to be impactful, formative, and even hurtful if not communicated with compassion. Furthermore, academic scientists are not often used to interacting with others from the place of the heart; instead, we privilege primarily intellectual interactions, even though emotional and psychological interactions are often just as relevant or important.

- Because, perhaps more than any other profession, professors are in a position to exert tremendous influence on students' and colleagues' futures, both in terms of their work habits, their career advancement, and their abilities to achieve harmony. This is a tremendous responsibility, one best tempered with the wisdom-kindness inherent in compassion.
- Because, in order to avoid bitterness and burnout, we must learn to be compassionate with ourselves, which in turn allows us to be compassionate to others. Developing compassion can lead to us discovering or rekindling our sense of joy, and to ensuring our longevity and sustainability as successful professionals.

This leads us to the five compassion practices we believe to be essential to meaningfully integrate compassion into our work and home lives. Like Singer, we believe compassion is a trait than can be cultivated through awareness and practice. If your compassion "muscles" aren't as developed as you would like, fear not! These are areas we can all work on continuously, and they are areas where we can improve, often with great results. The five practices are these:

(1) Opening your heart
(2) Being present
(3) Recognizing interdependence
(4) Drawing on wisdom
(5) Practicing kindness.

Exercise: one conscious act of compassion per day

We encourage you to view this exercise as an experiment that you can do with yourself. This experiment does not need to take much time, but it might shift your perspective in the way you interact with others. We suggest you consciously choose to do at least one act of compassion each day. This could be a small random act of kindness, such as letting somebody go ahead in traffic, or it could be a larger act, such as spending 30 minutes with a student who is struggling. There is, of course, no reason to do only one act of compassion; nothing keeps you from doing more. The point is to observe what happens when you consistently put your mind to doing one act of compassion per day. Such observation may be most effective when you write down each act of compassion that you carried out, along with the results of that act. Perhaps people shared something with you that you did not expect, or perhaps you received more compassion from others, or perhaps you simply started to view people differently or enjoyed your work more.

We have written about some of these practices in other chapters. But we expand on them here from within the context of compassion, and offer some practical tips for implementing them in your personal and professional lives.

OPENING YOUR HEART

The idea of opening your heart probably sounds ineffable or intangible, even a bit *mushy*. But we see it very much along the lines of the practices we discussed in Chapter 5 on Listening. Opening up the heart, in the professional or personal context, simply means taking the time to see or hear something from the perspective of caring for another, rather than from a perspective of trying to manage or control someone else, protect ourselves, or get something done. Opening the heart means seeing or sensing the inner person in ourselves or in others. Opening the heart requires that we slow down, make sure we have heard, understood, or seen the other from multiple perspectives, and then respond from a place of love. The watchwords for opening the heart are, *Is there something I might not be seeing, hearing, or understanding?*

Jen in particular has struggled with this compassion practice. Quick to respond, judge, or make sense of people and situations, she often assumes her perspective to be the "right" one. Sometimes these impulses serve us well – it can be important to trust our first impressions or intuition, and as academics we often succeed on the basis of quick thinking and analyzing. But slowing down and asking if there is something we may have missed, or another way of seeing things, can be a helpful counterweight to these impulses in our interpersonal interactions. Jen has learned over time that while first impressions or intuitions are very helpful tools, it is also wise and compassionate to remember that we often do not know the whole story. Often situations are more complicated and deserving of our love and understanding than we first perceive. In other words, we often know right away what our minds think of situations – our brains are incredible processing machines. But asking what our *heart* sees or hears can open us up to compassion.

A beautiful example of compassion from one of our interviewees, which we introduced in Chapter 5, is exemplary of the practice of opening the heart. You may remember that, having lost his own son to suicide, our colleague was able to "see" a struggling student in a new light. This student had a spate of automatic Fs on his record as a result of a battle with depression. Our colleague told us, "In evaluating his transcript and seeing the disastrous impact on his GPA of so many automatic Fs, I realized this student had dug a hole for himself that would take forever to remedy." Our colleague, whose heart had been opened by his own tremendous personal loss, was able to see something in this student that was not immediately visible, and advocated on his behalf to get him a clean slate. He notes, "I was thrilled at this outcome. It gave the student a new outlook on life and freed him to move in a new direction without a huge burden of academic failure."

It would have been so easy for our colleague to simply understand this situation from a place of intellectual rationality. This rationality might have encouraged him to reprimand this student for his "failures"; if we see students as often unwilling to work hard, or as not "making the grade," we might assume they are slacking off or are ill-equipped to succeed. Yet stern disciplinary action for this student – a "tough love" approach – probably would have backfired. Such an approach would have fed into this student's self-defeating belief system about his ability to succeed in graduate school.

It also would have been easy for this colleague, overwhelmed himself in a new position, undoubtedly concerned with dozens of important tasks, to not address the student's problems at all.

The student could have easily disappeared from the program, one way or another, without anyone ever intervening on his behalf. But because our colleague wondered if there was something he wasn't seeing – some explanation for the student's behavior beyond what was visible on a transcript – he was able to advocate for a compassionate solution to the student's problems. Our colleague wondered, *Is there something I might not be seeing, hearing, or understanding?* And this allowed him to act from the heart.

BEING PRESENT

"Being present" sounds deceptively simple. After all, how can we be anywhere other than where we are? Yet it might be the most challenging and difficult of all the practices we have introduced in this book. In general, Americans (and surely people from other cultures as well) struggle with *just being*. In fact, recent research suggests that silent "doing-nothingness" is so uncomfortable for many people that they will actually cause themselves physical pain to avoid it.

Struggling with just *being*.

A team of social psychologists in the United States set out to test subjects' ability to withstand periods of time in which they would have to do nothing other than sit and think (Wilson et al., 2014). During these "thinking periods," some individuals were allowed to think about whatever was on their minds, while others were given prompts, including thinking about what they would like to eat, to guide their thinking. Wilson and coworkers write that "on average, participants did not enjoy the experience very much: 49 percent reported enjoyment that was at or below the midpoint of the scale." In other words, half of us find *being present*, alone with our thoughts, or with nothing to do, either unpleasant or boring. Most of us wish we had more time and space in our lives, and yet when we have it, the experience is deeply uncomfortable.

The next phase of the research project delivered even more interesting results. In this phase, rather than simply measure tolerance levels for the "thinking periods," the team gave study participants the option of administering electric shocks *to themselves* in lieu of simply sitting in quiet. According to Wilson and his colleagues, "What is striking is that simply being alone with their thoughts for 15 minutes was apparently so aversive that it drove many participants to self-administer an electric shock that they had earlier said they would pay to avoid."

Clearly, not all of us would rather inflict pain on ourselves than just sit quietly, and there are undoubtedly reasons why sitting still for an experiment would be less fulfilling than sitting in our favorite park, relaxing. Nonetheless, we find the results of this study fascinating, because it suggests that there are many of us who really struggle with experiencing *what is*, with being present to whatever moment we find ourselves in. Yet the practice of coming back to the present moment is what allows us to connect with ourselves and with others as part of cultivating compassion.

In Chapter 3, we talked about the practice of meditation. We both have seen incredible long-term gains in our peace of mind and our ability to practice resilience and calm in difficult situations as a result of consistent meditation practice. Our guess is that if the study participants described earlier had a meditation practice, or something similar, they might have been less likely to shock themselves rather than be alone with their thoughts. Meditation practice also encourages us to be less reactive in particular situations – meditation sometimes allows us to develop skills that teach us how to sit with uncomfortable thoughts or feelings rather than react unfavorably or run away from them. This mindset of patience, and willingness to experience mental discomfort, rather than jump into problem-solving mode, or past the problem altogether, is also fruitful for cultivating compassion.

But what about the short term? What about situations where we are taxed, or struggle to maintain calm for any variety of reasons? Or we see a student or colleague who has reached a breaking point, or isn't meeting our expectations, or is just plain annoying, inappropriate, or angry themselves? How are we to locate compassion in such moments?

One particularly useful tool that assists in such situations is encapsulated in the acronym HALT. HALT is useful for situations ranging from dealing with toddlers having a tantrum, to relationship disputes, to working with angry faculty members. HALT stands for *Hungry, Angry, Lonely, Tired*. When we find ourselves, our loved ones, our coworkers, or our students in any one of these states, it becomes incredibly important to demonstrate loving-kindness, or compassion, to ourselves and to them.

Hungry. For example, if we are not in the habit of eating at regular intervals, or eating foods that provide us with sustenance and energy, we are much more likely to snap at colleagues and students. Parents of young children might be familiar with the saying, "Put your own oxygen mask on first." If you are hungry and having a low blood sugar yourself, you will be much less likely to be able to manage cheerfully pulling together a dinner for everyone else. If you are finding yourself feeling irritable and unable to manage things as well as usual, you might ask yourself if you are taking good care of your physical needs. If not, make some small changes to bring things back into harmony. We can extend the same understanding to others as well; perhaps one of our colleagues or students is having trouble caring for themselves properly and behaviors we find "annoying" are really a byproduct of them being out of harmony themselves. Perhaps, in other words, your difficulties with them have nothing to do with how they feel about *you*, and everything to do with how they feel about *themselves*.

"Hunger" can be metaphorical as well as spiritual. Perhaps we are hungry for beauty or quiet space or time in nature or play or creative expression. Pay attention to the longings of your body and your heart and see if addressing one of these areas helps with problems in your own life. Jen often uses the image of a balloon to describe this. If she is putting all of her energies into her work, it is as if she is squeezing one end of a balloon, tighter and tighter. Eventually there is no air left in the "work" end of the balloon, and furthermore, she feels like something elsewhere is about to explode! Clearly, this is not about work going poorly but about her releasing her tight grip over her work life and maybe giving some of the other areas of her life the space to breathe.

Angry. This is a tough one. We are not psychologists, and although we may recognize that someone in our lives is angry (and we know why, too), experience tells us that our friends and colleagues don't particularly like being psychoanalyzed by us! Many of us have complex reactions to anger as well. If someone is angry with us, or around us, we may become defensive, desire to flee, want to fix the problem, or get angry ourselves. We also believe that it can be very counterproductive to get in the *habit of anger*; some people we know seem to always be angry, and frequently anger leads to more anger. Remember that what we focus on tends to grow (Chapter 5). Furthermore, it is never okay to be abusive to others in the workplace, nor should aggressive or violent behavior be tolerated. At the same time, there are totally appropriate situations and contexts for anger;

anger at injustice, for example, can be used productively to address very problematic inequities in society. There are times and places where anger can play a powerful and formative role in moving us forward.

However, what we're talking about here is primarily the little flare-ups we have to deal with on a day-to-day basis in our personal lives and workplaces. Perhaps we have a student who is angry about a grade, or an administrator who is annoyed by a funding request, or a colleague who is jealous. We find it is helpful in these circumstances to remember that this kind of anger is often fueled by *fear*. The "grade-grubbing" student might have a demanding parent who expects straight As, for example. This student is afraid of losing his parent's love and acceptance. Or the administrator may be dealing with a major budget crunch not of her own making, and that she cannot share with you, and while she wants to give you what you ask for, she cannot. This administrator is afraid of seeming unfair or is

Remember that anger is often fueled by *fear*.

afraid of your anger toward her or is afraid she does not like her job very much at these difficult times. Our jealous colleague is afraid that he is not as smart or as good as you are, and that there will not be enough recognition to go around. Our colleague is afraid he is a failure.

We all have these fears, too. Because being afraid – and sharing that fear with others – makes us vulnerable, we often mask that fear as anger. Anger is, strangely enough, more culturally acceptable and easier to express than fear is. We have found the academic workplace to be particularly poor at supporting vulnerability; we are often much better at posturing or hiding behind our intellects than sharing our deepest fears and concerns with one another. In any case, it can be a helpful exercise when we are angry, or we encounter someone else's anger, to remember that the anger is probably coming from a fearful place inside, and that we can best take care of ourselves by identifying that fear, and interrogating what kinds of hidden beliefs or inner dialogues might be supporting it.

Roel was once upset and angry with a colleague. This colleague at one point simply asked, "What is it that you are really afraid of?" This question dissolved Roel's anger, and a fruitful and constructive conversation was the result. Sometimes we can assist each other in shifting the mindset away from anger by simply asking what our fear really is.

Lonely. University campuses are teeming with people and activities. This is part of what we love about the academic life. Although the arrival of the fall semester often arrives too quickly and we don't always feel ready to begin a new academic year, it is also exciting to know we will be meeting new students and colleagues, learning new things, and having new experiences. On even the smallest or quietest of campuses there are talks, film screenings, social gatherings, and – at the very least – meetings to attend.

Yet despite all the potential for socializing and learning and being with others, despite all the hustle and bustle, campuses and other scientific or academic environments can be, paradoxically, incredibly lonely places. Just as we can be at a loud party, crammed full of people, and feel totally alone, so too can we be in a vibrant, busy workplace and find ourselves feeling isolated or on our own.

Perhaps occasional bouts of loneliness are a part of the human condition. And we certainly don't mean to critique *solitude*, which could be thought of as the loving cousin of loneliness. Solitude is a consciously chosen time by ourselves, time taken to rest and recharge, to choose silence and contemplation over external chatter and expectations. Loneliness, on the other hand, is the sense that we are fundamentally separate from the humans around us. Loneliness invites us to compare ourselves unfavorably with others – we look around and believe others are having more fun or are more successful or feel more connected than we do. Such feelings of loneliness can lead to sadness, despair, and depression, or to unhealthy behaviors such as food, alcohol, and drug addictions or self-neglect.

A recent article published in *The New York Times* illustrates the ways in which experiences of loneliness are at once totally banal and quite pernicious. In the article, journalist Julie Scelfo writes about the experiences of students on college campuses who suffer tremendous feelings of loneliness and inadequacy, even as they present polished, happy images of themselves to the world via social media (Scelfo, 2015). The sense of loneliness, alienation, and impossible expectations is so great that some students choose to end their lives rather than continue to suffer. The phenomenon seems to be happening at several campuses: "In 2003, Duke jolted academe with a report describing how its female students felt pressure to be 'effortlessly perfect': smart, accomplished, fit, beautiful and popular, all without visible effort. At Stanford, it's called the Duck Syndrome. A duck appears to glide calmly across the water, while beneath the surface it frantically, relentlessly paddles" (Scelfo, 2015).

Faculty are not immune to such pressures, either. Female faculty – particularly those with families to care for – are often asked, "how they do it all." Academic "superstars" are those who are able to be excellent teachers, outstanding researchers, and great colleagues. We aim to be liked by our students, admired by our colleagues, and loved by our families. The sense that we must keep it all together – all while our feet relentlessly paddle under the surface of the water to keep us afloat – is a familiar sensation to many in the academy. Add to that the many pressures to defund public universities, eliminate tenure, or otherwise critique public school teachers and professors, and faculty can find themselves feeling overwhelmed and lonely indeed.

The point here is that, despite appearances, many of us struggle with strong feelings of being separate and alone. Noticing when we are ourselves feeling left out, sad, or threatened can be a very helpful form of self-knowledge. This is because feelings of loneliness can be addressed. While we may not all have (or want) the ideal romantic or life partner, or may struggle to make friends, to find "our people" or "tribe," we can find ways to meaningfully connect with others such that we don't feel so alone. Asking others questions, taking the time to listen to their answers, making an effort to participate in or organize social activities, or simply having honest conversations with others – even if it makes us feel vulnerable – are all ways to fight the pull of loneliness.

And, as was the case with hunger and anger, when we are kind to our lonely selves, we can also extend compassion to others who might be lonely. Perhaps a grumpy graduate student is struggling with feeling left out of his cohort, or an administrative assistant is going through a difficult divorce. Without violating personal boundaries, you could experiment with ways of reminding them they are a valued part of your work community. Graduate students can be invited to participate in reading groups where they discuss materials over potluck. Administrative assistants and colleagues can be asked out for group lunches. If they don't want to come, they will decline. But at least you will have made the effort to remind them they belong.

Tired. Tired is, in many ways, a contemporary sign of the times. Magazine headlines tell us that Americans don't get enough sleep, that we don't know how to rest, and that new and ever-expanding technologies are actually making it harder for us to sleep enough at night, turn off our brains, and enjoy uncontaminated time. It seems that many of us are chronically exhausted, a condition that can lead to all sorts of medical, psychological, and spiritual pressures on our health.

When we are not giving our bodies the rest they need, we may not be able to respond to or interact with others from a place of kindness or compassion. Pushing ourselves to work hard even when we are exhausted is itself a sign that we lack compassion for our own bodies and minds. Yet, there will still be times when we are tired and must nonetheless show up to do our jobs – whether that be teaching, advising, or working on our research. However, acknowledging that we are tired, off our game in the classroom, unclear in our communication with others, or temporarily ineffective in the lab, can itself be a form of compassion. We don't necessarily need to *complain* about our exhaustion, but cutting ourselves some slack at times when we are not functioning as well as we would were we fully rested is a sign of health and self-care. Then making sure to find the rest we need should become a priority.

When we value regular rest rather than relentless work guised as "productivity," we are also being good role models for those who observe our behavior, whether they be junior colleagues, students, or staff. We are also giving ourselves a chance to be even more effective in the workplace because we will be able to approach problems and projects with intellectual and physical freshness.

In sum, when we practice being present in these ways – by noticing if we are hungry, angry, lonely, or tired – we can often take care of ourselves in meaningful ways, ways that permit us to rest and recharge so that we can return to our responsibilities in a better frame of mind. Another beautiful side effect of approaching self-care in this way is that when you extend this kind of compassion to yourself, you can also more easily extend it to others. It is difficult, if not impossible, to be truly compassionate toward others in any lasting, sustained way, if we do not also show ourselves the same courtesy.

Exercise: develop antidotes to HALT

In this exercise, we encourage you to develop antidotes to hunger, anger, loneliness, and tiredness. Such antidotes might involve items you keep with you or in your office, or habits that you develop.

To stave off hunger you might keep some fruit or granola bars in your office. To quell anger, you might ask what you are, or somebody else is, really afraid of. To dispel loneliness, you could regularly take a colleague out for coffee or lunch. To be less tired, you could set a time by which you, or your graduate students, leave the office. These are just a few examples. Think about other ideas that are effective in countering HALT. Make these ideas practical and action-oriented, and make sure they work for you, so that they are easy to apply. Find out what works, what does not work, and use trial and error to develop habits that are effective countermeasures for hunger, anger, loneliness, and tiredness in yourself and those around you.

RECOGNIZING INTERDEPENDENCE

In the earlier sections, we argued that having compassion for ourselves often leads us to grant compassion to others. But we should also address why being compassionate with others matters. Compassion matters because it is an integral element in the web that connects us to each other as humans. When we forget to see each other as both connected and dependent on one another, we revert to a mechanistic style of interaction in which we believe ourselves to be alone, or we believe others are primarily there to serve or antagonize us. Forgetting that we are interdependent leads us to see competition instead of collaboration, isolation instead of connection, and manipulation instead of kindness. It is for these reasons that we return to the concept of interdependence that we introduced in Chapter 3.

We noted at the beginning of this chapter that as professors or leaders in our fields, we have a special responsibility to our students, colleagues, and coworkers. We are, fundamentally, in a "people" profession. Figuring out how to effectively work with and care for people is not some set of "soft skills" or add-on to our technical abilities. It is the centerpiece of a joyful life. As Mike Dooley puts it, "In all tests of character ... when two viewpoints are pitted against each other, in the final analysis, the thing that will strike you the most is not who was right or wrong, strong or weak, wise or foolish ... but who went to the greater length in considering the other's perspective" (Dooley, 2015). We talked about the phenomenon of the duck syndrome, wherein we make every effort to seem polished and effortless in our endeavors, while hiding our struggles and difficulties. The duck syndrome is evidence that we believe we cannot show our true selves to others, for fear of being judged or ostracized, and that we must instead maintain a façade of perfection and ease, even if that is not what we truly experience.

But allowing ourselves to be vulnerable – and to ask for help when we need it – is actually one of the *best* ways to connect with others in our shared humanity, if done so in a relationship where there is trust and mutual respect. An analogy might illustrate what we mean here: imagine that you had to spend every meal you ate, for the rest of your life, at a fine restaurant. Every meal would be perfection – beautifully prepared, both delicious and fulfilling, and no dirty dishes to contend with afterwards! But you would have to eat that perfect meal by yourself, with no human interaction. You would have no hand in preparing your food, and no one with which to discuss your day while you ate. This might sound lovely for a while (again, no dishes!) but soon you might grow a bit lonely. You might hunger for interaction with other humans. You might long to chop an onion, to have a say in the menu for the evening, or to set the table as you liked.

Conversation is loud and raucous, disrupted by peals of laughter.

Contrast this scenario with a regularly scheduled potluck. Perhaps eight of your dearest friends and family members show up at your house with all manner of drinks and dishes. The plates are mismatched, maybe your house isn't as clean as you would like, and the food is a mishmash of flavors and styles. Conversation is loud and raucous, disrupted by peals of laughter and the clanking of silverware. Your friends help you to tidy your kitchen before they leave, and you collapse on the sofa afterward, tired but happy, mulling over the conversations and joy of the evening.

We are guessing some of you might still choose our lonely restaurant scenario. But we believe that life – and our most real experiences of joy – happens in the potluck scenario. In the potluck scenario, people might see our messy homes or experience our less-than-perfect cooking or they might disagree with one another or leave a mess. But when everyone brings their gifts to the table, as imperfect as they may be, we have a chance to experience one another's humanity, and to become more at ease with our own. So it is with interdependence. We cannot get by on "perfect" very long, if it leads us to feelings of isolation and separation. We must find ways to remind ourselves of our deepest human connections with one another, and to explore what joy there is to be found in our honesty and imperfections.

Exercise: is your work like a fine restaurant or a potluck dinner?

We used the analogy of either having dinner alone in a fine restaurant or potluck dinner with friends to illustrate the value of interdependence. How does this apply to your work? Do you feel isolated, or do you feel embedded in a community? Does work serve you in a comfortable but lonely way, or is it an experience of give and take with others? Is your work joyful? At the end of the workday, are you full of the events and encounters of the day, of new impressions and ideas, or do you return home feeling isolated after another day in the office? Does your work feel sterile, or is it alive in a pleasingly messy way? Take a few minutes to think about these questions, and write down your responses. Based on these responses, is there anything you want to change? Is there anything you crave for? What steps can you take to bring about the change that you want or need?

DRAWING ON WISDOM

All this talk of compassion can raise questions and areas of concern. We ourselves discussed these at length while writing this chapter. For example, Jen, who has worked in male-dominated environments for many years, was concerned about how to best encourage compassionate behavior in those who want to become kinder in professional environments but who also have issues with perceiving boundaries. For example, we need to pay special attention to inclinations male professors might have to being paternalistic to female colleagues or students. Asking, "Aren't you just hungry or tired, sweetheart?" may seem loving or compassionate but in fact can be felt as condescending and inappropriate, depending on who is saying those words, who hears them, and in what context.

In other words, effectively practicing compassion requires some skill, knowledge, and self-reflection – what we think of collectively as "wisdom." Compassion requires that we be curious about the relationships in our lives but also that we question our role in creating and supporting structures of privilege, power, and access in the settings in which we work.

An example, Roel's "hugging dilemma," may help to illustrate. Roel is from the Netherlands, where physical affection – especially among men – is not typical. After decades of living in the United States, after developing close relationships with many of his coworkers and students, and after working with firefighters who interact differently than your average academic, Roel became more comfortable with giving and receiving hugs when the situation and context made them seem appropriate. And let's be clear – Roel doesn't go around hugging people willy-nilly. He uses wisdom and experience to figure out when it makes sense to offer or receive a hug! So, on the one hand, this ability to hug people he cared about in professional settings allowed Roel to express his sense of connectedness and interdependence with others. And others came to see him as "huggable," as well – someone who could be approached and counted on to lend a listening ear and a compassionate heart.

But now, a dilemma: Roel will hug people with whom he has a great relationship, and it is natural and easy to do so. The hugging "seal," so to speak, has been broken. What about when Roel meets new people, though? Does he offer a hug? Is it strange to hug one colleague and offer the other a handshake? Are there times when Roel should not accept hugs, say from a student whom he is about to evaluate in a qualifying exam? Is it ever acceptable to hug a female colleague or a student behind closed doors? What if Roel hugs a male colleague and that colleague is clearly very uncomfortable being hugged? Should he apologize, laugh it off, move on?

This example probably seems a little absurd – all this talk of hugging! But we bring it up because it's a good example of one of those gray areas where we might wonder if we are acting "professional" enough, but also acting from a place of the heart. One of our colleagues, who is in a position of leadership, is very physical: she puts her arm around shoulders and liberally gives hugs. When asked why she does this, she replied, "These engineers live so much in their left brain; physical contact draws them back to their right brains." Indeed, physical contact may help foster warm and balanced interactions. But scientists might want to preserve boundaries so as to protect themselves and their students, while also maintaining a rich level of connection and kindness. Also, we may have been "burned" in the past when we misread a situation or overstepped boundaries and now we have retreated into strict rules of conduct that prohibit us from meaningfully interacting with those around us. Figuring out how to move forward from that place of regret or shame is tricky.

We can't provide a manual here for when and how to physically express affection or connection any more than we can provide an in-depth manual for when and how to "act compassionately" in the office setting. We can only say that there is some trial and error involved, and that you must develop some wisdom around how to reach out when compassionate responses are called for, particularly if you do not feel you are someone for whom these kinds of interactions come naturally.

Roel encountered an interesting example of such questions about boundaries in action, when he developed a course on research ethics with his colleague Carl Mitcham, who is an interdisciplinary scholar specializing in ethics. After they taught the class, Carl thanked Roel profusely for co-teaching the class. When asked why he was so thankful, Carl answered, "You made it so personal, by continually asking what the issue we were discussing meant for the personal and professional life of the students in the class." Roel suggested to Carl that both of them could do more of this, but Carl said that he couldn't. As a professor, he had reservations about encouraging his students to express personal values: in the classroom this might be perceived as undue pressure. More importantly, he felt that since ethics is a field of research for him, he had to keep an emotional distance. This puzzled Roel because he is aware of Carl's deep passion for the topic of ethics, and especially for teaching ethics. But the episode illustrates the inherent tension between the personal involvement, passion, and connections we bring to our students and the classroom, and the objectivity and boundaries that we often seek to maintain in our teaching and research.

If either of these dilemmas – the hugging dilemma or the classroom dilemma – resonates with you, then you may need to do some more work around what it means to access compassion for yourself and others, and then to express it appropriately. When in doubt, we say begin with yourself. If you are unsure of how to properly express compassion to others, do more work extending compassion to yourself. One place to start is with what Buddhist practitioners call the "loving kindness meditation." Sit in a comfortable place where it is quiet and you won't be disturbed. If it feels good, place a hand over your heart. Then you can repeat these sentences to yourself, breathing in and out with each phrase:

> *May I be safe.*
> *May I be peaceful.*
> *May I be kind to myself.*
> *May I accept myself as I am*
> *(Singer & Bolz, 2013).*

It is remarkable how much inner wisdom and guidance we can access when we ourselves begin from a place of peace and self-kindness. Taking time to be centered can also give you the time you need to act appropriately and compassionately in different situations. Again, as academics and reasoned, rational professionals, we are often trained to act and respond quickly. But as writer and entrepreneur Danielle Laporte puts it, "Hesitation can be a form of wisdom. Motives become clearer; new information shows up. Amazing grace can happen when you choose inner rhythms over external pressure" (Laporte, 2015).

PRACTICING KINDNESS

These many steps culminate in the act of *practicing* kindness. This is the doing of compassion. We assume that many of you already practice kindness, and much of this chapter feels familiar to you. If, however, you are like us and have had to develop a compassion practice over the years, or you are just starting to think about how to integrate compassion into your personal and professional life, then this final step is where the rubber meets the road.

Simply put, practicing kindness requires that we act not from a place of shame, guilt, or anxiety but from a place of acceptance, honesty, and peace. We find that academics are often trapped by rules they set for themselves – rules about deadlines, or the meaning of success, or busy-ness. We have encouraged you to question these assumptions or rules throughout this book and to instead discover what is true for you. If the rules you live your professional life by come from a place of shame, guilt, or anxiety, they are most likely not going to lead you to joy, and they almost certainly will not lead you to practice compassion and kindness for yourself and others. On the other hand, if you can decide to see many options before you – if you can loosen the rules you have set for yourself somewhat – you may find that opportunities for joy, kindness, and compassion appear. As Martha Beck puts it, "Liberation is the ability to see multiple options. No matter how silly a circumstance may sound, if it gives you the ability to widen your perspective, it helps set you free" (Beck, 2009, p. 63).

Once we understand that there are multiple ways to engage and connect with those around us, we find that there are a host of small actions we can begin to take in order to practice kindness on a daily basis. And we encourage readers to begin with small actions – it can be

We can take the time to notice when someone is suffering and needs a kind word or space or even an embrace.

overwhelming to think about "practicing compassion" on a large scale. Not all of us are ready to become philanthropists or begin nonprofit organizations. Luckily, compassion can happen on a small scale, through our daily actions with one another. We can choose to act with kindness rather than react with anger. We can forgive ourselves when we make mistakes. We can offer leniency instead of punishment. We can use wisdom to establish boundaries that also allow us to see each other from a place of the heart. We can take the time to notice when someone is suffering and needs a kind word or space or even an embrace.

We love this excerpt from the 1953 poem "The Pleasure of Serving," by Gabriela Mistral (translated here by Gloria Garafulich-Grabois):

All of nature is anxious to serve.
Clouds serve. The wind serves. The groove serves.
Where there is a tree that needs to be planted,
    be the one to do it.
Where a task requires the effort everyone avoids,
    be the one to do it.
Be the one who will move the stone off the road
    between hate in the hearts and the difficulties of the problems.
There is happiness in being good and just.

But above all, there is the beautiful, the immense happiness of serving.
How sad would the world be if everything was done,
    if there was not a rosebush to be planted, a new enterprise to start!
Do not only be tempted by the easy tasks.
    It is so wonderful to do the tasks that others avoid!
But do not make the mistake of thinking that the merit is obtained
    only by doing great tasks.
There are small tasks that are good tasks: arranging a table,
    organizing a home, combing a child's hair.

Let others be the ones to criticize. The ones that destroy.
    You be the one that serves.
Serving is not a task done only by inferiors.
God, who gives the fruit and light, serves.
We could say: He who serves and who has his eyes
    fixed on our hands, asks us each day:
—Who will you serve today? A tree, your friend, your mother?

Mistral was the first Latin American woman to receive the Nobel Prize in Literature, and she was a prolific poet, writer, and activist who traveled widely and worked internationally. In short, she was influential, and she was busy! But in this poem she reminds us that in each moment, we have the opportunity to choose how we see and interpret our lives and our situations. When tempted to listen to our inner critic, we can instead choose to listen to our inner protector or guide. We can choose kindness instead of anger, connection instead of loneliness, and rest instead of exhaustion. We can feed the many hungers of our bodies and souls, and create opportunities for others to do the same.

    Those are the fruits of compassion.

7   Integrity

We aren't all called to be a Mandela or a Gandhi. But if we pay
attention, we will find that life calls to us every day to go beyond our
own interests. And, when we do, our own interests are served in ways
that are inexplicably and profoundly meaningful.

–Alex Pattakos, *Prisoners of Our Thoughts* (2008)

Having read this book, you probably could have guessed that we
believe a core attribute of the joyful scientist is integrity. If we cannot
be honest with ourselves and others in a compassionate and caring
way, it will feel nearly impossible to be at home in our own skins and
to encourage others to also feel free to express, explore, and learn. For
us, therefore, integrity is the ability to *be ourselves regardless of the
context*. Of course, we encourage you to pay attention to the mores
and expectations of the workplace, and discourage you from dishing
out the "honest truth" in order to make yourself feel righteous at the
expense of others. But we do think that finding ways to bring our
whole selves into the workplace, and paying attention to harmony in
all parts of our lives – in other words, *integrating* the different pieces of
ourselves, rather than segmenting or compartmentalizing – is
a characteristic of those scientists who seem to have the joy thing
figured out.

In this chapter, therefore, we encourage you to reflect both on
how you might integrate the practices of harmony, courage, vision,
curiosity, and compassion into your own life. We suggest you make
a plan to try a few new things to shift the flow of energy and attention
in your own life. Some of us are capable of making big changes
overnight. For most of us, however, we must take small steps,
adopting new micro-practices a little at a time until they become
new habits. When we are used to seeing the world as working one
particular way, it can take time to shift our view.

What do you integrate these topics with? First of all, these topics can be integrated with each other. For example, your vision may depend on your willingness and ability to be curious or to listen, your courage may allow you to take your vision to a higher level, and compassion may help to realize your vision in alignment with the needs of others. In such ways, the topics of the previous chapters are connected, and it is in such connections that these topics become alive. But second, we encourage you to integrate the material of this book into your daily life. We encourage you to write in the margins of this book, photocopy pages, rip them out and tape them up, and otherwise thoroughly *use* the book in whatever way works for you.

Exercise: articulate one step to practice the habits described in this book

Hopefully this book inspires you and gives you ideas. But to integrate the habits we describe, it helps to reflect on ways in which you can apply the habits we describe on a regular basis, perhaps a daily one. In order to make this specific we suggest you finish the following sentences in a way that is meaningful and practical for you.

- In order to make my life more harmonious, I will ...
- To be more courageous, I will...
- I will develop a greater vision by ...
- I will grow and express my curiosity by ...
- To listen better I will ...
- I will express more compassion by ...

There is, of course, no reason why you should stick to one idea or action item for each of the aforementioned habits. But we encourage you not to be too ambitious, because this may hamper the implementation of your intentions.

One should be able to see how a bridge carries a load.

WHAT DO WE MEAN WITH INTEGRITY?

Roel's father, Wim, was aesthetic advisor for the Netherlands department of transportation and waterworks. He was involved in designing bridges, and did so with great passion. In his view, crossing a bridge had to convey the sense of being part of a dramatic event in which one crosses from one bank of the river to the next. He felt strongly that one should be able to see how a bridge carried a load. And from a construction point of view, he wanted the load to be carried and distributed in the most elegant way, a way in which all the bridge's elements would collaborate to carry the load on the bridge and to carry the structure itself. All these elements – the different structural elements of the bridge, the balance of forces that supports the load, the people who use the bridge, and the experience of using the bridge – are the *integrity* of the bridge in the sense that all of this is integrated into what the bridge is and what it means to those who use the bridge.

The metaphor of building such a bridge is a good one for thinking about building personal integrity. When we live in personal integrity, the different activities and roles in our lives merge in a harmonious way into what we are, what we do, and what we mean for the world around us. In this view, integrity points to a type of wholeness where the different parts of our life are in harmony with each other and with the world around us. When we live like this, our life is whole. The wholeness of integrity means that we live a life without compartments (Palmer, 2004). When we stand in life in such an integrated way, we acknowledge that we have different roles to play in different situations, but that these roles are not placed in separate boxes, as pieces to be managed; they are all integrated into our own nature. The integration of roles into one personality is something that may take attention, perhaps even effort. In an academic environment we likely are a teacher and/or researcher. But we also are children, we may be parents, we may be a member of a sports club, political organization, or we may do volunteer work. In these different situations we play different roles, but when doing this in integrity these roles express in different ways the one person that we really are. When this expression is harmonious, there is no conflict or contradiction between these roles; they simply are different facets of our life.

In practice these roles do not always align with each other. A professor who is devoted to his children might be impatient and careless with his students (or the other way around). In which of these situations do we see the real face of the professor? Does this professor know what his real face is? This opposite of integrity has been called "incoherence" by the world-renowned physicist David Bohm (1996, pp. 88–89). He writes about this state of mind: "Incoherence means that your intentions and your results do not agree. Your action is not in agreement with what you expect. You have contradiction, confusion, and you have self-deception in order to cover it up." A lack of integrity comes with the price of contradiction (we act in different ways in different situations), confusion (who is the real "me"?), self-deception (it is natural that I act in such different ways), and cover-up (I did not really mean to do this). Living with contradiction, confusion, self-deception, and cover-up can take up enormous mental energy, which may create a sense of feeling stressed and drained. And there is a more insidious part of this behavior: our life is internally divided.

This division can be confusing, both for ourselves and for those around us.

Exercise: in what aspects of your life do you sense incoherence?

Our lives have many different facets, and the degree to which these different facets are well integrated may vary. The definition of incoherence – a misalignment of intentions and results – can be useful for discovering where we can bring greater harmony in our life. What are the areas in your life where your intentions and the results are different? What may cause this discrepancy? What could you change to bring your intentions closer to their results? Think of at least one action item to create more coherence in your personal or professional life. Write down this action item and post it on a place where you encounter it regularly, and observe what happens when you follow up on this action item.

Integrity implies being present and assuming responsibility. This word becomes clearer when we write responsibility as "response-ability"; we respond fully and consciously to the here and now. It is our ability to respond meaningfully while being present that allows us to make the best of our situation. It is in attention to the here and now that our ability to choose stares us in the face, because it is in the present that we can decide what sort of presence we choose to be. We cannot force other people to change. Also, we can neither change who we were (the past) nor who we are going to be (the future). Therefore our span of control really lies within ourselves and in the present moment. It is only in the choice of our own thoughts, words, and actions right where we are that we can make a difference. This choice can affect others and it may change our future, but the choosing is done for ourselves and it is done in the present. That does not mean, of course, that our choices have no bearing on the past or the future. For example, the choice of forgiveness frees us from burdens from the past. Likewise, the choices that we make today may change our future. But the choosing itself is always done in the present.

Ultimately integrity is related to harmony. When we live in integrity the different voices and roles in our life combine into something greater. Each of these voices contributes to our life, and it is in integrity that these contributions are harmonious. Are the different parts of what you are and what you do in harmony or in dissonance? In Chapter 5 on Listening we asked that you listen to the harmony in your life, or the lack thereof. Another way of asking this is to encourage you to tune in to the degree of integrity in your life.

Integrity brings simplicity. When we live in integrity, our words, values, and actions are aligned. This means that we are honest and that we live by our values. We speak the truth, and when we do so we don't need to remember what we said to different persons or in different situations because our thoughts, words, and actions are consistent with each other and consistent in different situations. In this way of living there is no game playing, no backstabbing, and no politicking. Isn't that a great improvement over a life where we have to make sure certain facts do not see the light of day, where we manipulate, and where our life is a political game? Imagine how simple committee meetings would be if they were held in integrity! There would, of course, still be different views and interests – dissent is a great contributor to wise decision-making – but there would be a transparency and breadth of goodwill that would make such dissent much easier to handle.

In integrity you only have one face. In the story "The Strange Case of Dr. Jekyll and Mr. Hyde" by Robert Louis Stevenson, the same man appears either as a pleasant sociable citizen, Dr. Jekyll, or as a monstrous sociopath, Mr. Hyde.[1] The book is a parody about a split personality – the antithesis of integrity. In the story, the transformation between Jekyll and Hyde is involuntary, and is on several

---

[1] It is amusing to note that the pleasant personality, Dr. Jekyll, holds an advanced academic degree while the horrible Mr. Hyde does not.

occasions influenced by a potion. The involuntary aspects of the transformation between Jekyll and Hyde makes us aware how easy it is to have multiple faces. Indeed, it can take an effort to live in integrity, particularly if you are not used to doing so, or if it feels easier to dissemble or act one way around some people, and differently around others. On the other hand, when we change our face in different situations, we need to remember which face to carry, and who has seen the different aspects of us. Once we get used to living in integrity, however, we find that the energy it takes to be ourselves across context is much less than the energy it takes to try to be different things to different people. In fact, there is nothing more draining, and nothing that robs us of our authenticity and joy more quickly.

Living in integrity means that we truly participate in life. The concept of participating is described eloquently by Bohm (1996, pp. 98–99), who writes that

> [participation] really has two meanings. The earliest meaning was "to partake of," as you partake of food – people all eating from a common bowl, partaking of bread or whatever it is. . . . The second meaning is "to partake in," to make your contribution . . . Taken together, these two ways of thinking do not create a separation between object and subject.

Participation viewed in this way is a way of life in which giving and taking are integrated. As Bohm describes, the view of participation as integrated giving and taking ultimately makes the distinction between object and subject – the giver and the taker – disappear. Participation implies a level of sharing that is the foundation of living in community with others, whether it is at home, at work, or otherwise. It is the idea of participation in community that takes integrity from a personal and individual concept to that of coherence and collaboration of a group of people. Integrity thus is both a personal and communal concept.

Exercise: living in community?

In this exercise we ask that you reflect on the degree to which you truly participate, as described by David Bohm, in your professional life. The following questions may help you to do so.

- Do you feel that your work is a shared commitment and opportunity for expression with your colleagues, or do you carry the burdens and reap the fruits in isolation?
- What do your contribute to your work environment? How do others contribute to your endeavors? Are your efforts, and those of your colleagues, shared, or are they focused on individual achievement and advancement?
- Does your work environment feel like a community where you care for each other, or it is more like a group of individuals who happen to work at the same place?
- If your work does not have a sense of community, would you like to change it? And if so, what steps could you take to create a larger sense of community? Are you willing to take one or more of these steps?

GUARDING INTEGRITY

The description of integrity given here hopefully is inspiring. But in reality, living in integrity requires attention, focus, and courage. We have seen in the story of Jekyll and Hyde that changes in character can happen involuntarily, and that it may take an effort to prevent such changes. We may need to stand up at times for our integrity since the expectations of other people, our employer, or situations that we may find ourselves in may not be aligned with who we are and what we want to be.

A prerequisite for living a life of integrity is that we pay attention. If we live on autopilot and go through the day automatically, we don't act, but instead merely react. This reactive, perhaps instinctive, response does not necessarily express who we choose to be. Living in integrity requires conscious choice, it means switching off the autopilot. Such switching off is exactly what is meant by the words "paying attention." And as stated by Alex Pattakos in the quote at the start of this chapter, when we pay attention to life, we begin to discover life. We discover depth. We discover meaning.

It thus requires focus to live with integrity. The busy-ness of the modern work environment can easily divert us and take us away from the life that we seek to live. We described in Chapter 3 the importance of formulating a vision. In practice it requires focus, attention, and intention to follow through on our goals and vision in the face of the demands that we encounter. Living in integrity implies that we take our goals and vision seriously, that we focus on them, that we guard them, that we stand up for them.

In some situations, courage is needed to stay the course. Courage may be needed to stand up for our goals and vision. In Chapter 2 we discussed sensitive conversation topics in the academic environment. It may take courage to speak about such topics when they are important to us. Courage may be needed to address an injustice. It may require courage to turn down a research grant because it has strings attached that are not in line with our values. It may require courage to show our colleagues or our students that we care about them, because we might think that such an "emotional" attitude is not appropriate at work. And it even may require courage to say simple things such as "I am sorry" or "I really appreciate what you do." Such simple words are sometimes hard to express, especially in a "brainy" environment. Yet it is through such simple words that deep connections can be forged.

INTEGRATE IN YOUR WORK ENVIRONMENT

Many of us feel that we have to go through our studies and our career all by ourselves. Our educational system promotes individual achievement. Every student receives an individual grade for each test, and these are combined into an individual Grade Point Average, a number that students carry with them and that is often used as a summary of their achievements. This continues after our formal education. As scientists we each have our individual publication

Every student has an individual Grade Point Average, a number that students carry with them and that is often used as a summary of their achievements.

record and impact score, and receive our personal annual evaluation. In a company we might receive our personal bonus. It is no wonder that after having been trained in a system with such an individualistic approach to measure success, we believe that we have to achieve this success all by ourselves. Yet the reality of being an effective academic is that we collaborate. We connect. This often comes naturally to many of us when it concerns scientific projects or teaching. But do we turn to others when we have doubts and question the type of work that we do? Do we seek input from others when things don't go well? Or do we stay quiet, fearing that others will judge us for being weak or incapable?

Here is the good news: many, if not most, people are willing to help once we ask. This may seem counterintuitive and risky. Many of us have heard stories of or have experiences with exploitative or manipulative people in our workplace. Maybe we ourselves have not always acted as helpfully as we could. But we would remind readers about the "mean world" hypothesis – if you are looking for those who are out to do you in, you will probably find them. If you really have an issue with a coworker or advisor who is intentionally or thoughtlessly destructive, then you must seek help for that and act to address that problem. But we would argue that in most cases most people are not only willing to help you but they would love to help you. The role of helper usually is gratifying because we can use our expertise or personality in a positive and constructive way. But help usually doesn't arrive spontaneously; we must ask for it. Asking for help can be the hardest part of the process, but once we have gone over that bump, the input, expertise, and support from others usually becomes available to us.

Ideally, the academic environment is set up to provide you with opportunities to reach out and get help when you need it. During our education we have hopefully had an advisor who listened to us, mentored us, and counseled us in wise and productive ways. After our formal education ends, the advisor usually disappears from our life, with the result that we are on our own. But that needn't be so, and shouldn't be. It really helps us throughout our career when we have a mentor or support system. Some universities formally assign a mentor to young faculty members. But you don't need to wait for such a formal assignment. In fact, mentorship does not need to be formalized to be effective. The most important criterion for a mentor is that you and your mentor operate on a basis of trust and mutual appreciation, and that your mentor has the experience that you need. So get yourself a mentor! Roel, while in his fifties, was seen by many students and peers as an experienced and senior-level professor. But Ken Larner, who is about 20 years older, has been an informal mentor to Roel himself in many different ways: by being available to answer questions or discuss dilemmas, by sharing his wisdom and expertise, and by a balanced combination of challenging and giving moral support.

Jen, on the other hand, has had her most fulfilling academic relationships with peers who have worked as a support system – support that has helped her through her most vulnerable professional and personal moments. While at a conference several years ago, Jen met three colleagues from universities all around the United States. They agreed to get a drink one night after panels, and (after a few beers) all four revealed that they felt a bit stuck in their writing, frustrated with the academic grind, and yearning for work to feel fun and exciting again. As the night progressed, the four began to brainstorm ideas for how they might work together in the coming months, and they filled a few cocktail napkins with project ideas.

It would have been very easy to leave this connection at that – as a scribbled bar napkin, or to chalk it up to blowing off steam after a long day of panel connections. But instead, the four kept in email contact over the following weeks, and eventually committed four days that summer to meeting at a cabin in the middle of the United States and trying to write an article together. Several years later, Jen and her colleagues still attend conferences together (staying in a rented condo instead of the conference hotel), have peer-reviewed publications in print, have won several awards, and have written a book together.

But none of that would have happened had the four colleagues not expressed on that first night their doubts about writing, their fears of feeling isolated, and their desire for something more.

Sometimes, you have to ask to get what you want. You have to listen to your inner voice to know what you want, and then you need to be willing to listen when the answers come to you, sometimes in forms you don't expect. Your willingness and ability to listen is also essential in your interactions with those that you, in turn, support. These will be your students, and perhaps also colleagues. The degree to which you listen to your students may be related to the way you see students, as we discussed in Chapter 4 on Curiosity. When you see students as nameless units that need to be filled with knowledge, you likely have a completely different interaction with them than when you see them as individuals with their own skills, charms, quirks, and challenges.

Exercise: seek help in one area where you need it

Asking for help comes naturally for some of us. If that is the case, you can skip this exercise. But if you are one of the many people for whom asking for help is difficult, we suggest that you identify one area of your life where you could benefit from some help.

We suggest you choose the area where you have the greatest resistance to seek help, because this is where you might benefit most. Who could you ask for help? And what is it that you exactly ask for? After you have answered these questions, be bold and ask for the help that you need. How was your request for help received? What happened when you followed up on the help that was given to you?

TO WRAP IT ALL UP

We end this book with the chapter on integrity because we want to encourage you to integrate harmony, courage, vision, curiosity, listening, and compassion in your personality and in your daily life. Integrating these different aspects of our personality is an interesting challenge. The integration of our personality in the deepest sense with the way we lead our life at work, at home, at the gym, at parties, or elsewhere can be even more challenging. The reality is that we are full of contradictions: between emotional detachment and personal involvement in our work, between love of self and altruism, and in many other ways. These contradictions express the richness of our character and our life. The challenge is to weave these contradictions into the tapestry that is the "us." This effort is worthwhile, though, because integration gives a depth and meaning that greatly enriches our life. This depth and meaning is not just valuable and rewarding for ourselves; when we experience depth and meaning in ourselves, we automatically bring it forth to others. What happens when our integration brings forth depth and meaning is beautifully described in the following words by Clarissa Pinkola

Estés (1996): "When a man gives his whole heart, he becomes an amazing force – he becomes an inspiratrice ... He carries the seeds of new life and necessary deaths. He inspires new work within himself, but also in those near him."[2]

Roel received a beautiful inspiration to live in integrity from his father. Roel always wanted to play the soprano saxophone, but waited until his mid-thirties before he actually started playing. In fact, he did not start playing the soprano saxophone. Instead, he purchased an alto saxophone, because everybody told him that the soprano was so much more difficult to play. Around the same time, Roel's father was diagnosed with cancer, and he knew he would not live long. One night the bell rang at Roel's home and his father was at the doorstep to deliver the following message: "If you want to play the soprano saxophone, why don't you do it today? Why postpone until tomorrow what you can do today? I want to give you a soprano saxophone." The instrument that he gave Roel became a precious parting gift from his father. But more importantly, what a gift it was for Roel to receive the wise lesson to seize the day and live in the moment in such a powerful way. This is the type of gift that we make when we become an inspiration for others.

We all have the power to be an inspiration, which means that YOU have the power to be the *one who inspires*. Every day we have opportunities to be a light and to inspire others. This book is really aimed at helping you to live an integrated life and to be a light and inspiration in an academic environment. We end this book with the following parting words that integrate the topics of this book in a heartfelt wish and encouragement.

> Have the courage to be your self and to shine your light. Create
> a vision of your greater yet to be, both for yourself and for those
> around you. Be curious to push forward, to explore the unknown,

---

[2] The word "death" is not meant in this quote as the end of physical life. It refers to the termination of activities or personality traits that no longer serve us.

and to go beyond current limitations. Listen and take in both the
external world and the internal world. Have compassion to be
a beneficial presence, and have the integrity to be harmonious
and whole.

Be well!

References

Anxiety and Depression Association of America. (2015). *Facts and Statistics*. Retrieved July 2, 2015, from adaa.org: www.adaa.org/about-adaa/press-room/facts-statistics.

Beck, M. (2003). *The Joy Diet: 10 Daily Practices for a Healthier Life*. New York, NY: Crown Publishers.

Beck, M. (2008). *The Four-Day Win: End Your Diet War and Achieve Thinner Peace*. Emmaus, PA: Rodale Books.

Beck, M. (2009). *Steering by Starlight: The Science and Magic of Finding Your Destiny*. Emmaus, PA: Rodale Books.

Bohm, D. (1996). *On Dialogue*. New York, NY: Routledge.

Brown, B. (2015). *Daring Greatly: How the Courage to be Vulnerable Transforms the Way We Live, Love, Parent, and Lead*. New York, NY: Gotham Books.

Cain, S. (2013). *Quiet: The Power of Introverts in a World that Cannot Stop Talking*. New York, NY: Random House.

Coelho, P. (2008). *The Witch of Portobello*. New York, NY: Harper Perennial.

Collins, H. M. & Pinch, T. (2012). *The Golem: What You Should Know about Science*. New York: Cambridge University Press.

Colvin, G. (2010). *Talent is Overrated: What Really Separates World-Class Performers from Everybody Else*. New York, NY: Penguin Books.

Covey, S. (1990). *The 7 Habits of Highly Effective People*. New York, NY: Fireside Books.

Csikszentmihalyi, M. (2008). *Flow: The Psychology of Optimal Experience*. New York: Harper Perennial Modern Classics.

Davis, R. (2014, October 22). *You at the Center of Your Life*. Retrieved from http://us9.campaign-archive1.com/?u=6b3196e6f73747c083721a156&id=f1bd72a804&e=c486a525b4 (accessed on February 25, 2016).

Dooley, M. (2015, July 23). *Notes from the Universe*. Retrieved July 23, 2015, from TUT: www.tut.com/Inspiration/nftu.

Estés, C. (1992). *Women Who Run with the Wolves*. New York, NY: Ballantine Books.

Estés, C. (2011). *Walking Strong In Two or More Worlds: Diversity and Solid Identity*. Lecture at the Colorado School of Mines.

Frayling, C. (2005). *Mad, Bad and Dangerous? The Scientist and the Cinema*. London: Reaktion Books.

Freeman, R. B. & Huang, W. (2014, February). *Collaborating With People Like Me: Ethnic co-authorship within the US*. Retrieved February 12, 2015, from The National Bureau of Economic Research: www.nber.org/papers/w19905.

Gardner, H. (2008). *5 Minds for the Future*. Cambridge, MA: Harvard Business Press.

Gauchat, G. (2012). Politicization of Science in the Public Sphere: A Study of Public Trust in the United States, 1974 to 2010. *American Sociological Review*, 77 (2), 167–187.

Gerber, L. (2014). *The Rise and Decline of Faculty Governance: Professionalization and the Modern American University*. Baltimore, MD: John Hopkins University Press.

Gerbner, G., Gross, L., Morgan, M., Signorielli, N., & Shanahan, J. (2002). Growing Up with Television: Cultivation Processes. In J. Bryant & D. Zillman (eds), *Media Effects: Advances in Theory and Research* (Second edn, pp. 43–68). Mahwah, NJ: Lawrence Erlbaum and Associates.

Goertzen, M. (2014, August). *What I'm Learning about Balance and the Simple Life*. Retrieved June 17, 2015, from Shalom Mama: http://shalommama.com/balance.

Goleman, D. (2004, January). What Makes a Leader? *Harvard Business Review*, 1–10.

Hammerskjöld, D. (1964). *Markings*. London, UK: Faber and Faber.

Heath, C. & Heath, D. (2007). *Made to Stick: Why Some Ideas Survive and Others Die*. New York: Random House.

Herkert, J. (2003). Microethics, Macroethics, and Professional Engineering Societies. *National Academy of Engineering*. Washington, DC: National Academies Press, pp. 107–114.

Katie, B. (2015). *Do the Work*. Retrieved June 25, 2015, from TheWork.com: http://thework.com/do-work.

Kuhn, T. (1962). *The Structure of Scientific Revolutions*. Chicago, IL: University of Chicago Press.

Kunnecke, A. (2015). *Annakunnecke.com*. Retrieved July 2, 2015, from www.annakunnecke.com.

Laporte, D. (2015, July 23). *#Truthbomb elaborations*. Retrieved July 23, 2015, from daniellelaporte.com: www.daniellelaporte.com/procrastination-can-be-a-form-of-intuition/?inf_contact_key=012f524628667987147b655d9b9139283ff2b86473d6c4e353007137d0f03c05.

Lombrozo, T. (2013, December 2). *The Truth about The Left Brain / Right Brain Relationship*. Retrieved July 6, 2015, from npr.org: www.npr.org/sections/13.7/2013/12/02/248089436/the-truth-about-the-left-brain-right-brain-relationship.

Mason, M. A., Wolfginger, N. H., & Goulden, M. (2013). *Do Babies Matter? Gender and Family in the Ivory Tower*. New Brunswick, NJ: Rutgers University Press.

Medin, D. L. & Lee, C. D. (2012, May/June). *Diversity Makes Better Science* (A. f. Science, Producer). Retrieved February 12, 2015, from Observer: www.psychologicalscience.org/index.php/publications/observer/2012/may-june-12/diversity-makes-better-science.html.

Medina, J. (2008). *Brain Rules; 12 Principles for Surviving and Thriving at Work, Home, and School*. Seattle, WA: Pear Press.

Melton, G. D. (2014). *Carry On, Warrior: The Power of Embracing Your Messy, Beautiful Life*. New York: Scribner.

Milan, L. M. & Hoffer, T. B. (2012, January). *Racial and Ethnic Diversity among U. S.-Educated Science, Engineering, and Health Doctorate Recipients: Methods of Reporting Diversity*. Retrieved February 12, 2015, from National Science Foundation InfoBrief: www.nsf.gov/statistics/infbrief/nsf12304/

Minsker, B. (2010). *The Joyful Professor. How to Shift from Surviving to Thriving in the Faculty Life*. Oconomowoc, WI: Maven Mark Books.

Mistral, G. (1953). The Pleasure of Serving. Translation by Gloria Garafulich-Grabois (c), published in *From Chile to the World: 70 years of Gabriela Mistral's Nobel Prize* (Gabriela Mistral Foundation, Inc.: New York, 2015). All rights reserved.

Mooney, C. (2005). *The Republican War on Science*. New York: Basic Books.

Mooney, C. & Kirshenbaum, S. (2010). *Unscientific America: How Scientific Illiteracy Threatens Our Future*. New York: Basic Books.

Moore, T. (1992). *Care of the Soul*. New York, NY: Harper.

Olson, R. (2009). *Don't Be Such a Scientist: Talking Substance in an Age of Style*. Washington, DC: Island Press.

Oreskes, N. & Conway, E. M. (2011). *Merchants of Doubt: How a Handful of Scientists Obscured the Truth on Issues from Tobacco Smoke to Global Warming*. New York: Bloomsbury Press.

Oxman, A., Chalmers, I., & Liberati, A. (2004). A Field Guide to Experts. *British Medical Journal*, 329, 1460–1462.

Palmer, P. (2004). *A Hidden Wholeness: The Journey Toward an Undivided Life*. Hoboken, NJ: Jossey-Bass.

Pattakos, A. (2008). *Prisoners of Our Thoughts*. New York, NY: Berrett-Koehler Publishers.

Prigogine, I. & Stengers, I. (1984). *Order Out of Chaos*. Toronto, Canada: Bantam Books.

Ray, S. A. (2013). *Amit Ray Quotes* (I. L. Publishers, Producer). Retrieved July 22, 2015, from Amit Ray: http://amitray.com/amitray_quotes/

Roberts-Miller, T. (2014, August 25). 9 to 5. Retrieved from Inside Higher Ed: https://www.insidehighered.com/advice/2014/08/25/essay-working-40-hours-week-academic (accessed on February 25, 2016).

Robinson, K. (2006, February). *How Schools Kill Creativity*. Retrieved from TED Lectures: www.ted.com/talks/ken_robinson_says_schools_kill_creativity?language=en

Robinson, S. (2012, March 14). *Bring Back the 40-Hour Work Week*. Retrieved February 12, 2015, from Salon.com: www.salon.com/2012/03/14/bring_back_the_40_hour_work_week/

Rosser, S. V. & Taylor, M. Z. (2009, May–June). *Why Are We Still Worried about Women in Science?* (Academe, Producer) Retrieved February 12, 2015, from American Association of University Professors: www.aaup.org/article/why-are-we-still-worried-about-women-science#.VN0dZkImXG4

Scelfo, J. (2015, July 27). *Campus Suicide and the Pressure of Perfection*. Retrieved July 29, 2015, from The New York Times: http://nyti.ms/1VIuROq

Schulte, B. (2014). *Overwhelmed: Work, Love, and Play When No One Has the Time*. New York: Sarah Crichton Books.

Sindermann, C. (1985). *The Joy of Science*. New York: Plenum Press.

Singer, T. & Bolz, M. (Eds.). (2013). *Compassion: Bridging Practice and Science* (Max Planck Society). Retrieved July 22, 2015, from http://www.compassion-training.org/en/online/index.html?iframe=true&width=100%&height=100%#2/z

Snieder, R. & Larner, K. (2009). *The Art of Being a Scientist*. Cambridge, UK: Cambridge University Press.

Stokstad, E. (2014). The Mountaintop Witness. *Science Magazine*, 343, 592–595.

Turk, T. (2009). *The Raven's Gift*. New York, NY: St Martin's Griffin.

van der Post, L. (1961). *The Heart of the Hunter*. London, UK: Penguin Books.

van der Post, L. (1992). *Interview with Sandra Harrington for the South Africa TV Channel*. Retrieved from www.youtube.com/watch?v=iQEsKzXUSxE

Wanjek, C. (2013, September 03). *Left Brain vs. Right: It's a Myth, Research Finds*. Retrieved July 6, 2015, from LiveScience: www.livescience.com/39373-left-brain-right-brain-myth.html

Wilson, T.D., D.A. Reinhard, E.C. Westgate, D.T. Gilbert, N. Ellerbeck, C. Hahn, C. L. Brown, & A. Shaked (2014). Just Think: The Challenges of the Disengaged Mind, *Science*, 345, 75–77.

Women in Global Science and Technology. (2012). *Scorecard on Gender Equality in the Knowledge Society* (T. a. National Assessments on Gender and Science, Producer). Retrieved February 12, 2015, from http://wisat.org/data/documents/GEKS_Scorecard-Highlights.pdf

Young, V. (2011). *The Secret Thoughts of Successful Women: Why Capable People Suffer from the Imposter Syndrome and How to Thrive in Spite of It*. New York, NY: Crown Business.